The British Horse Society
STAGE 2
WORKBOOK

A study and revision aid for exam candidates

Melissa Troup BA, BHSII and
Margaret Linington-Payne MA (Ed), BHSI

KENILWORTH PRESS

First published in the UK in 2008 by Kenilworth Press, an imprint of
Quiller Publishing Ltd

Reprinted 2012

British Library Cataloguing in Publication Data
A catalogue record for this book is available from the British Library

ISBN 978-1-905693-23-8

Layout and illustrations by Carole Vincer
Cover design and prelims by Sharyn Troughton

Printed in Malta by Gutenberg Press

KENILWORTH PRESS
An imprint of Quiller Publishing Ltd
Wykey House, Wykey, Shrewsbury
Shropshire, SY4 1JA
tel: 01939 261616 fax: 01939 261606
e-mail: info@quillerbooks.com
website: www.kenilworthpress.com

DISCLAIMER: The authors and publishers shall have neither liability
nor responsibility to any person or entity with respect to any loss
or damage caused or alleged to be caused directly or indirectly
by the information contained in this book. While the book is as
accurate as the authors can make it, there may be errors, omissions,
and inaccuracies.

CONTENTS

INTRODUCTION

This workbook has been compiled as a revision aid for candidates preparing for the BHS Stage 2 exam. It is designed to be used in conjunction with a Stage 2 course, ideally provided by a BHS Where to Train Centre, where instructors have a good understanding of the BHS examination system.

The questions have been written to captivate the imagination and help to make revision and quizzing of knowledge entertaining, whilst maintaining the integrity of the exam for which the student is preparing.

The authors wish to stress that there is no 'BHS way' for either practical or theory. As such there may be more answers to questions than have been given. The BHS system aims to train practical, safe and efficient horsemen and women, thus providing a foundation of internationally recognised qualifications from which a person may develop in any equestrian direction.

Details of further reading and contact details for the BHS are given at the end of the book.

BHS exam workbooks are available for Stage 1 and Stage 2

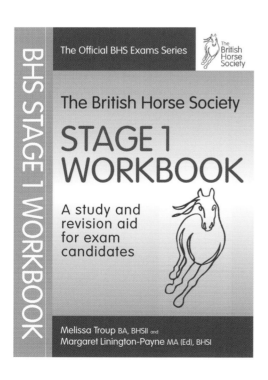

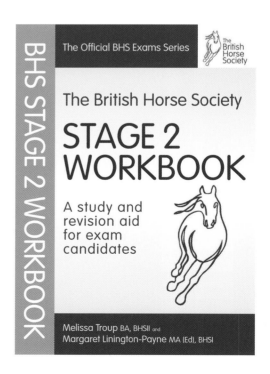

1 GROOMING

Q1.1 Complete the table to outline the differences between quartering, strapping and brushing off.

	QUARTERING	STRAPPING	BRUSHING OFF
HEAD		YES	
BODY	YES		
MANE AND TAIL	YES		
LEGS			
FEET PICKED OUT		YES	
FEET WASHED AND OILED			NO
EYES AND NOSE SPONGED			
DOCK SPONGED			
STABLE STAINS SPONGED			
STABLE RUBBER			NO
BANGING			
TIME TAKEN			

Q1.2 List five safety precautions for you and/or the horse when grooming.

1. _____

2. _____

3. _____

4. _____

5. _____

Q1.3 Why is it important to be efficient when grooming?

_____

_____

Q1.4 Imagine that you have been handed a hot, sweaty horse after an intense show-jumping training session. Using the key words below, complete the table to describe your after-work care.

TACK	
WALK	
WATER	
UNTACK AND WASH	
RUG	
WATER	
TACK	
INSPECT	
WATER	
GROOM AND RUG	

Q1.5 **(a) When plaiting, why do we use an odd number of plaits along the neck (making an even number when including the forelock)?**
(b) If the horse has a short or long neck, can this be helped visually and, if so, how?
(c) Describe how to plait a mane correctly, as shown in the drawings A – E.

(a) _____

_____

(b) _____

_____

_____

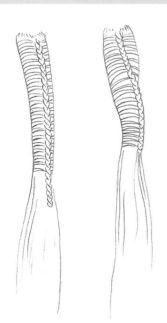

(c) A _____

B _____

C _____

D _____

E _____

Q1.6 **(a) What is the difference between a 'flat' and a 'ridge' tail plait?**
(b) Compare these two tail plaits. Write down why you think one is well plaited and the other poorly plaited.

(a) _____

_____

(b) _____

_____

_____

_____

_____

2 CLOTHING

LIGHTWEIGHT NZ

MIDDLEWEIGHT NZ

HEAVYWEIGHT NZ

SUMMER SHEET

LIGHTWEIGHT COOLER

HEAVYWEIGHT COOLER

LIGHTWEIGHT STABLE RUG

HEAVYWEIGHT STABLE RUG

STABLE NECK COVER

Q2.2 Correct blanket fitting. Put the following drawings into the correct order and describe each.

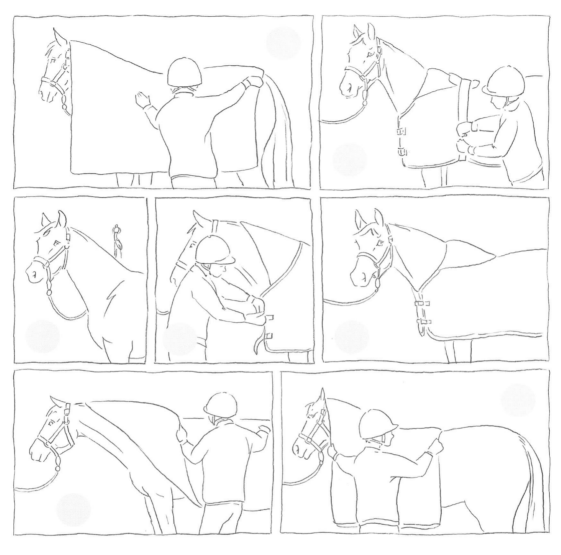

1 _____

2 _____

3 _____

4 _____

5 _____

6 _____

7 _____

_____

Q2.3 Give four reasons for using a stable bandage.

1. _____

2. _____

3. _____

4. _____

**Q2.4 This cartoon strip shows the correct procedure for applying a stable bandage.
Describe each picture.**

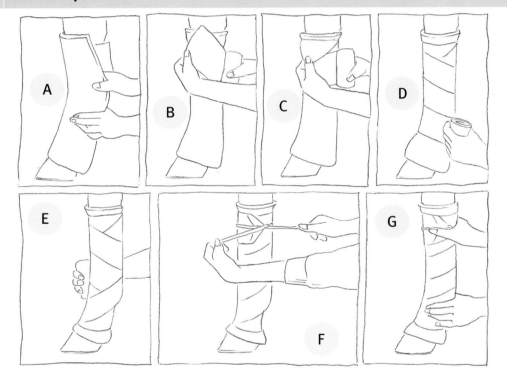

A _____

B _____

C _____

D _____

E _____

F _____

G _____

Q2.5 **What is the difference between Fybagee for stable bandages and Fybagee for travelling bandages?**

_____

_____

Q2.6 **List the equipment used for travelling and state the purpose of each item.**

_____

_____

_____

_____

_____

_____

_____

Q2.7 Fill in the gaps to explain how to apply and fit travel boots.

correctly	slippage	centre	strap	clean

Tie the horse up securely. On _____, dry legs fit each boot individually, positioning _____ and

then fastening. When fastening, start with the middle _____ and work up and down the leg from the

_____ out. Fasten firmly into position, ensuring no _____. Bandages are often

considered safer than boots.

Q2.8 What are the dangers of using travelling boots/bandages that are not correctly fitted?

_____

_____

_____

_____

Q2.9 How do you clean and store rugs, boots and bandages?

RUGS

_____

BOOTS

_____

BANDAGES

_____

FYBAGEE

_____

ALL

_____

3 SADDLERY

Q3.1 Looking at the image of this well-fitting saddle below, list the points to check when fitting a saddle.

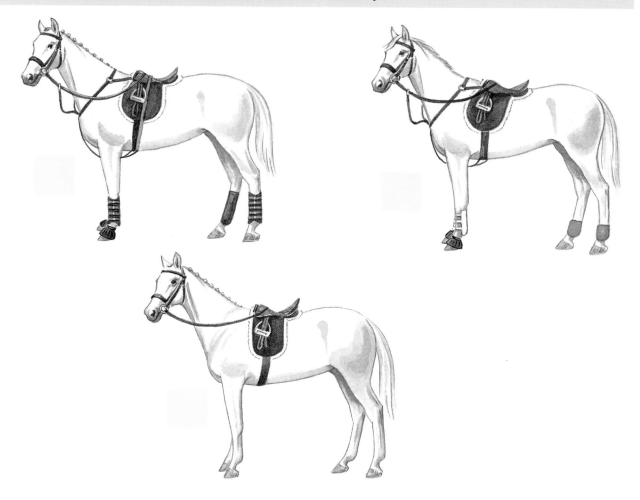

SIZE	
LEVEL	
POMMEL CLEARANCE	
GULLET	
SHOULDER	
BEARING SURFACE	

Q3.2 Look at these drawings and decide which would be suitable for:
(a) dressage   (b) show jumping   (c) cross-country

Q3.3 List five possible consequences for horse or rider if ill-fitting tack is used.

1. _____

2. _____

3. _____

4. _____

5. _____

Q3.4 Look at this horse tacked up for lungeing. Give an explanation of the fit of each item of equipment.

BRIDLE	
CAVESSON	
SADDLE	
SIDE REINS	
BOOTS	
LUNGE LINE	

Q3.5 Identify and give the function of each boot below.

BOOT	FUNCTION

Q3.6 **Looking at the images of the running and standing martingales below, describe the correct fit of both.**

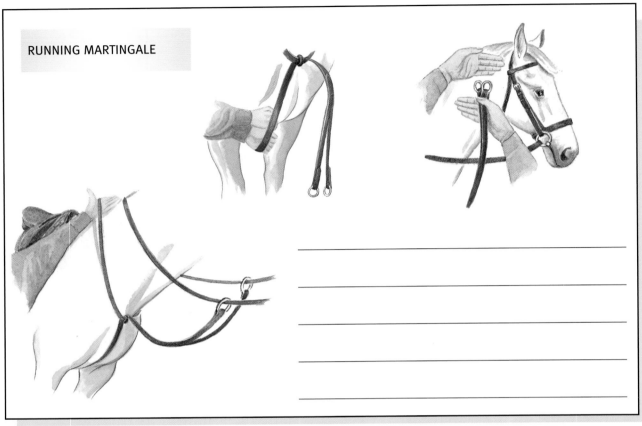

RUNNING MARTINGALE

_____

_____

_____

_____

_____

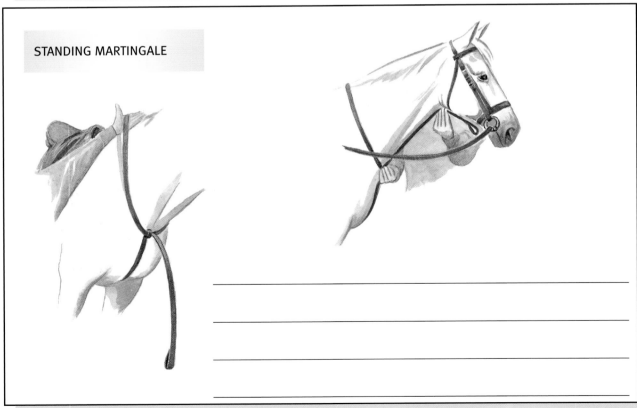

STANDING MARTINGALE

_____

_____

_____

_____

Q3.7 **What would you look for when checking the fit of a hunting breastplate whilst tacking up?**

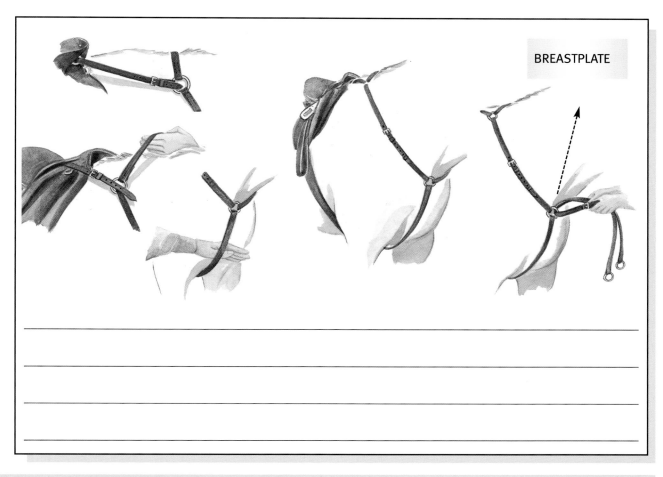

BREASTPLATE

_____

_____

_____

_____

_____

Q3.8 **Explain the purpose of each stage of tack cleaning.**

WARM WATER	
OIL	
SADDLE SOAP	

Q3.9 **Imagine that there are pieces of tack that need to be stored. Explain your preparation of the tack and how you would store it.**

_____

_____

_____

_____

4 HANDLING AND LUNGEING

RISK	MUCKING OUT	GROOMING	LEADING
SELF			
HORSE			
OTHERS			

Q4.2 **Circle the qualities that you feel would help a groom to maintain the confidence and control of horses.**

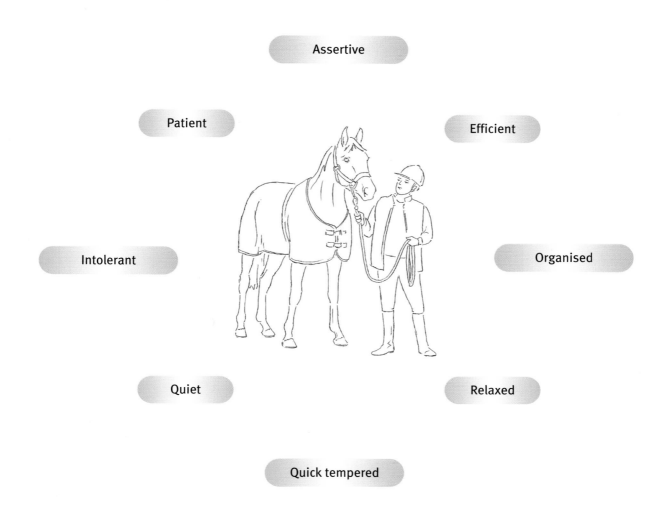

Assertive

Patient

Efficient

Intolerant

Organised

Quiet

Relaxed

Quick tempered

Q4.3 **List five reasons for lungeing.**

1. _____

2. _____

3. _____

4. _____

5. _____

Q4.4 The two people lungeing below show very different stances. One will appear confident to the horse, the other, not. List the differences in body language.

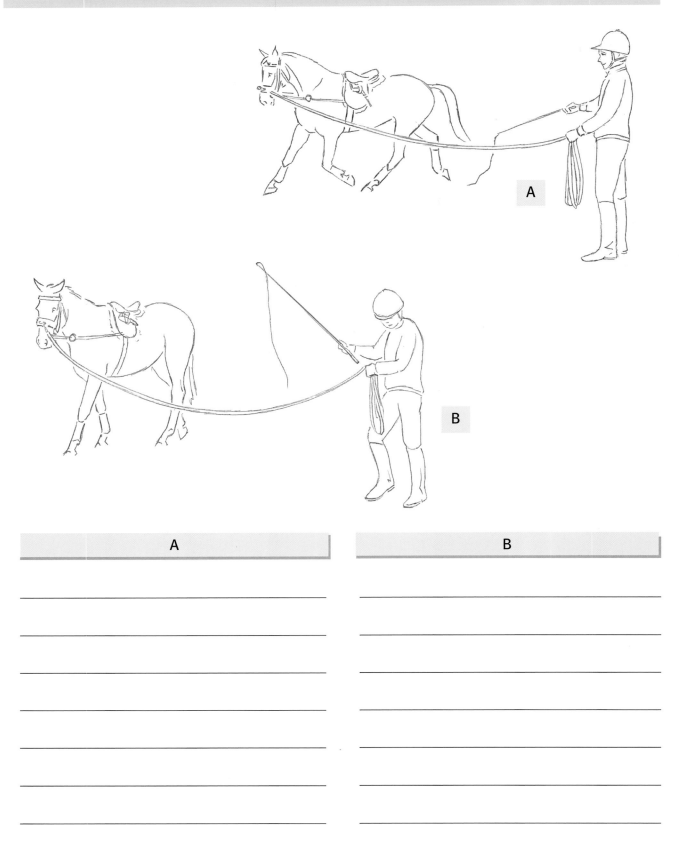

A
_____
_____
_____
_____
_____
_____
_____

B
_____
_____
_____
_____
_____
_____
_____

Q4.5 What 'feel' should you have at the end of the lunge line?

_____

Q4.6 List two main rules to follow when you are lungeing.

_____

_____

Q4.7 Explain the benefits of each of these when lungeing.

WALK	

TROT	

CIRCLE	

TRANSITIONS	

CANTER	

Q4.8 Give the reasons for using each part of the lungeing equipment.

CAVESSON	

SIDE REINS	

BRUSHING BOOTS	

LUNGE LINE	

SADDLE/ROLLER	

LUNGE WHIP	

Q4.9 Why must the handler wear a hat, gloves and correct footwear when lungeing?

5 STABLE DESIGN

Q5.1 List one advantage and disadvantage of each of the stable materials below.

	ADVANTAGE	DISADVANTAGE
WALLS		
Breeze block		
Brick		
Wood		
ROOF		
Tiles		
Plywood and felt		
Corrugated iron		
FLOOR		
Brick		
Concrete		
Dirt		

Q5.2 Complete the table of ideal stable dimensions.

	16 hh HORSE	PONY
SIZE		
HEIGHT		
DOOR WIDTH		
DOOR HEIGHT		
HALF DOOR HEIGHT		

Q5.3 (a) List three methods of ventilating stables.
(b) Why is ventilation in stables important?

(a) _____

_____

_____

(b) _____

_____

Q5.4 (a) **Why is good drainage necessary in stabling, and in which direction should the floor slope?**
(b) **What are the possible results of poor drainage?**

(a) _____

(b) _____

_____

_____

_____

_____

_____

Q5.5 Complete the table to include essential and extra stable fittings.

ESSENTIAL	EXTRA

Q5.6 Why should there be as few stable fittings as possible?

_____

6 SHOEING

A

B

C

D

E

F

G

H

I

J

K

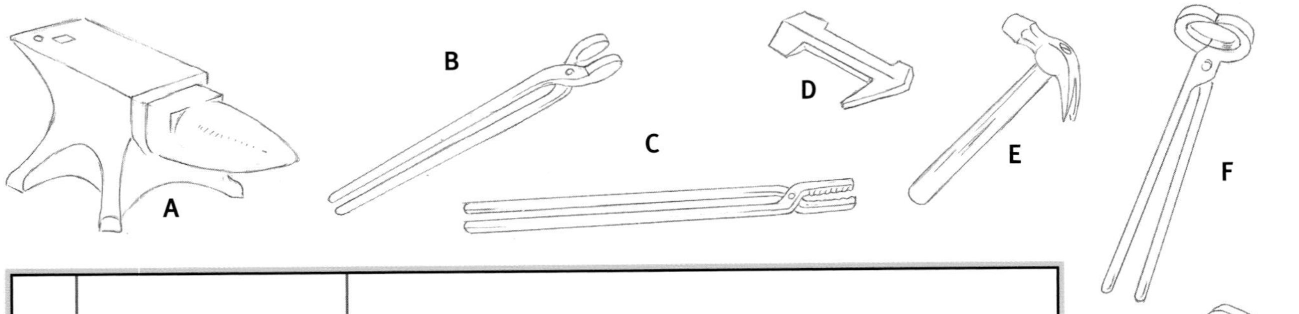

A		
B		
C		
D		
E		
F		
G		
H		
I		
J		
K		

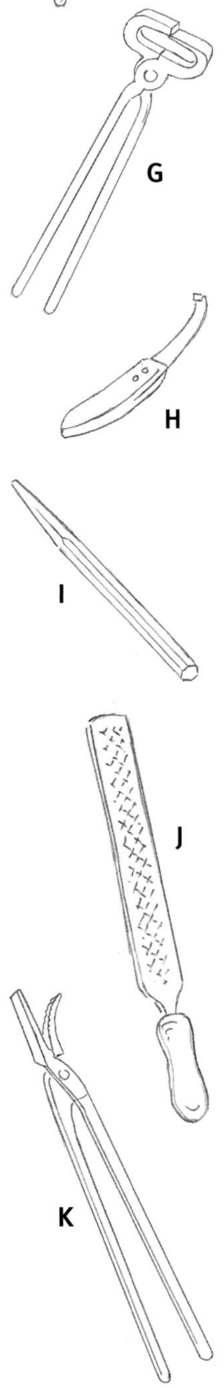

Q6.2 This cartoon strip outlines the shoeing procedure. Explain what is happening in each picture.

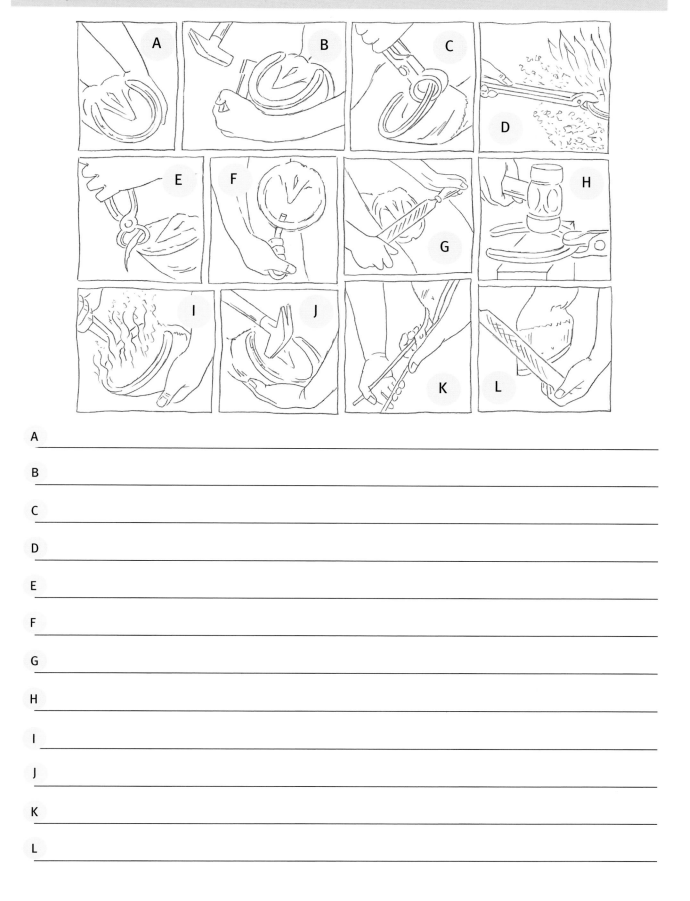

A _____

B _____

C _____

D _____

E _____

F _____

G _____

H _____

I _____

J _____

K _____

L _____

Q6.3 What procedure would you follow for removing a loose or twisted shoe?

_____

_____

_____

_____

_____

Q6.4 The pictures below show possible problems that can arise from infrequent shoeing. Identify each.

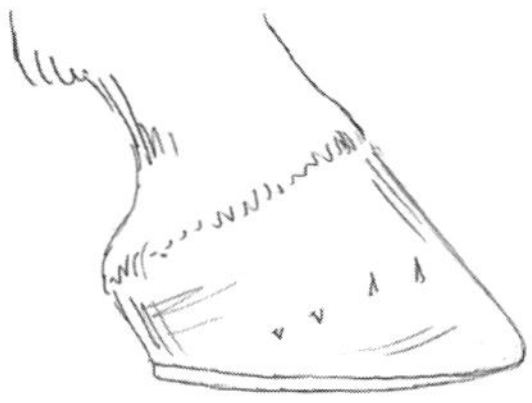

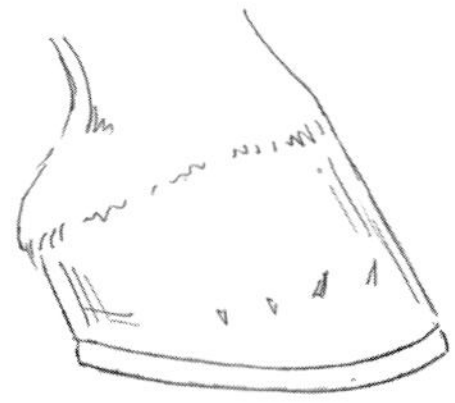

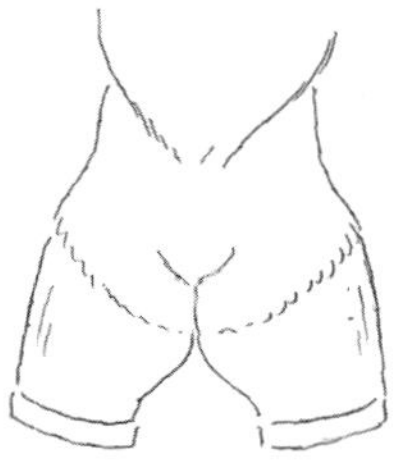

Q6.5 Explain the possible consequences of each of the problems in Q6.4.

Q6.6 **Below is a series of statements about what to look for in a newly shod foot. Label them TRUE or FALSE accordingly.**

Foot fits shoe

Clenches should be flush with the foot

The bearing surface is level

The farrier has chosen the correct type and weight of shoe for the horse

The horse is sound

The shoe fits the foot

The front feet look oval, the hind, round

Q6.7 Looking at the pictures, what problems may occur if unshod hooves are not trimmed regularly?

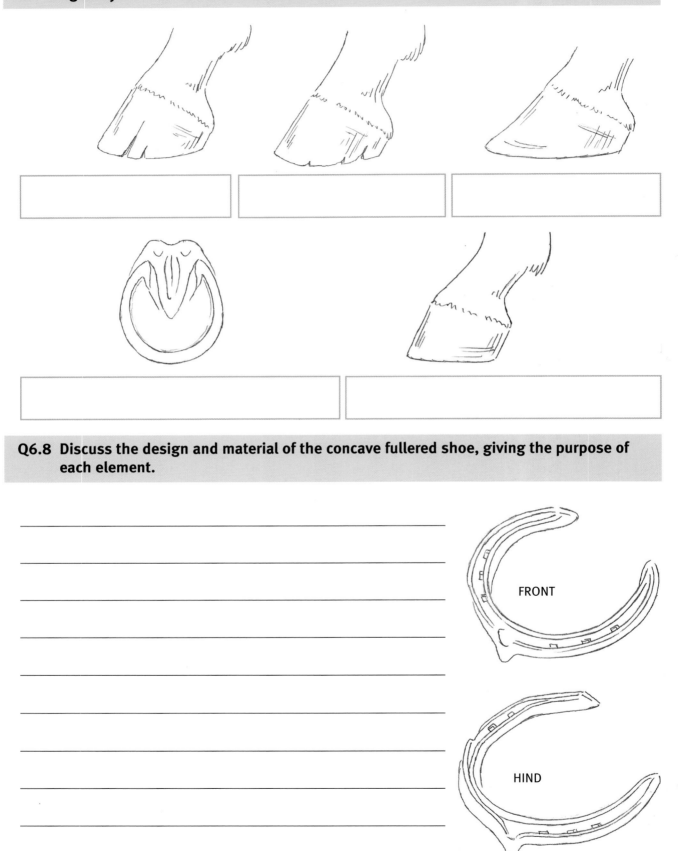

Q6.8 Discuss the design and material of the concave fullered shoe, giving the purpose of each element.

_____

_____

_____

_____

_____

_____

_____

_____

_____

FRONT

HIND

7 CLIPPING AND TRIMMING

Q7.1 Circle the correct reasons for pulling a mane.

Shows off the horse's neck

Makes the mane lie better

Thins the mane

Makes the mane thicker

Easier to plait

Makes the mane longer

Makes the mane less greasy

Q7.2 List two reasons why you would not trim a horse.

_____

_____

Q7.3 Looking at the horse below, label the areas that can be trimmed.

Q7.4 On the drawing in Q7.3, indicate whether you could use a comb (C), round-ended scissors (RS), or clippers (CL) for each of these areas. You may choose more than one for any area.

Q7.5 Give four reasons for clipping and explain the benefits of each.

1

2

3

4

Q7.6 Describe a full clip and a hunter clip.

Q7.7 Draw the appropriate clip on each horse.

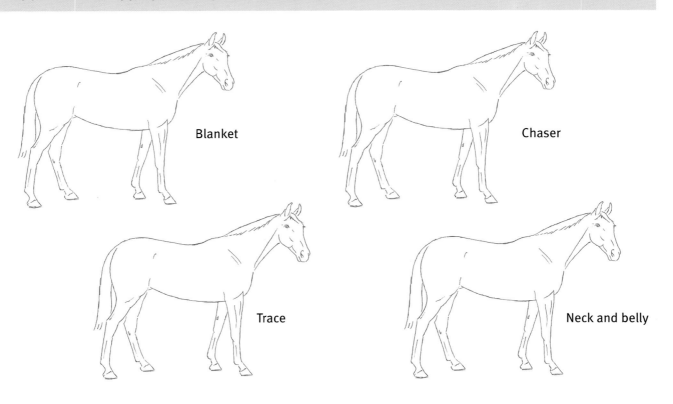

Blanket

Chaser

Trace

Neck and belly

Q7.8 What workload and lifestyle would be suitable for each of these clips?

CLIP	LIVE IN/OUT	AMOUNT OF WORK
BLANKET		
CHASER		
TRACE		
NECK AND BELLY		

Q7.9 Here is a picture of a pair of clippers, an extension cable and a circuit breaker. Label and annotate the drawings, indicating the routine checks and maintenance that are made before, during and after clipping.

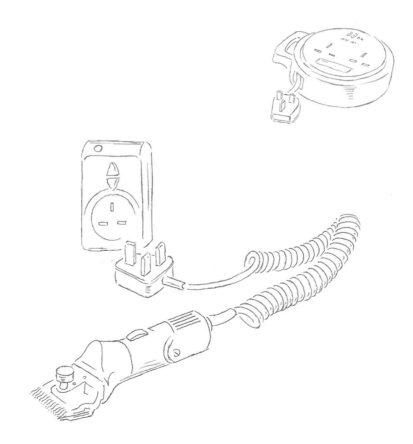

Q7.10 Complete the chart to list all aspects of health and safety for yourself, the horse and the handler when clipping.

YOU

HORSE

HANDLER

8 ANATOMY AND PHYSIOLOGY

Q8.1 This picture shows a healthy looking horse. List the signs of good health.

_____

_____

_____

_____

_____

_____

_____

_____

_____

_____

_____

Q8.2 On the diagram below, label the horse's internal organs.

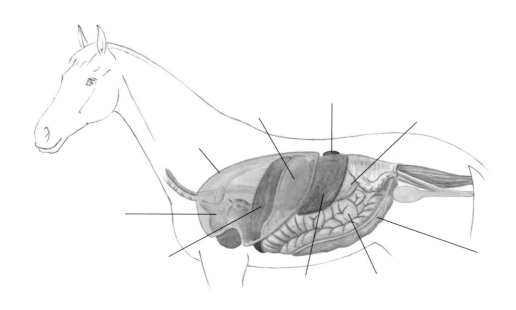

Q8.3 On this diagram of the digestive system, from mouth to small intestine, label each part, and then briefly describe its function.

A _____

B _____

C _____

D _____

E _____

F _____

G _____

H _____

I _____

J _____

K _____

_____

_____

_____

L _____

Q8.4 Here is a diagram of the second half of the digestive system. Label each part, and then briefly describe its function.

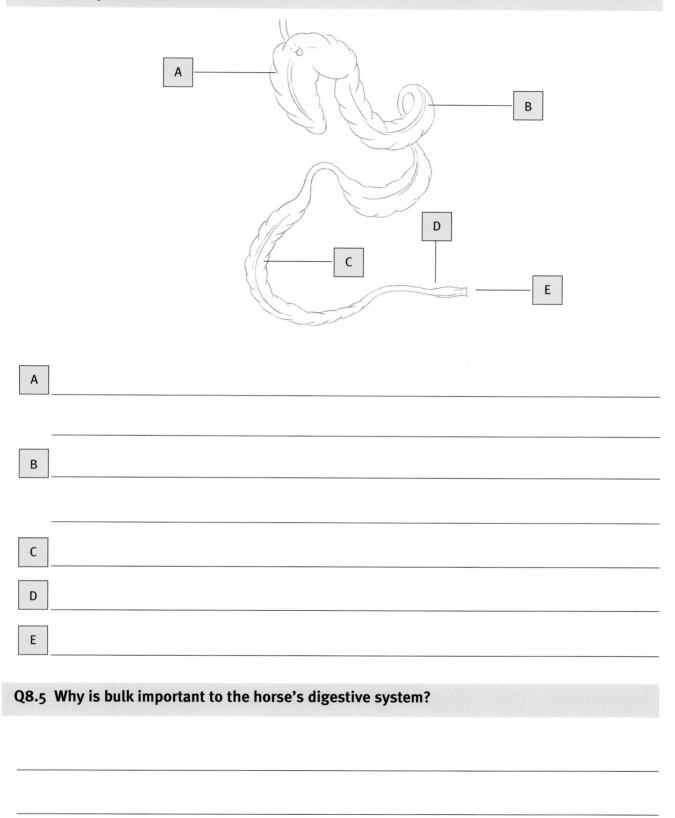

A _____

B _____

_____

C _____

D _____

E _____

Q8.5 Why is bulk important to the horse's digestive system?

_____

_____

Q8.6 Label the bones in the horse's skeleton.

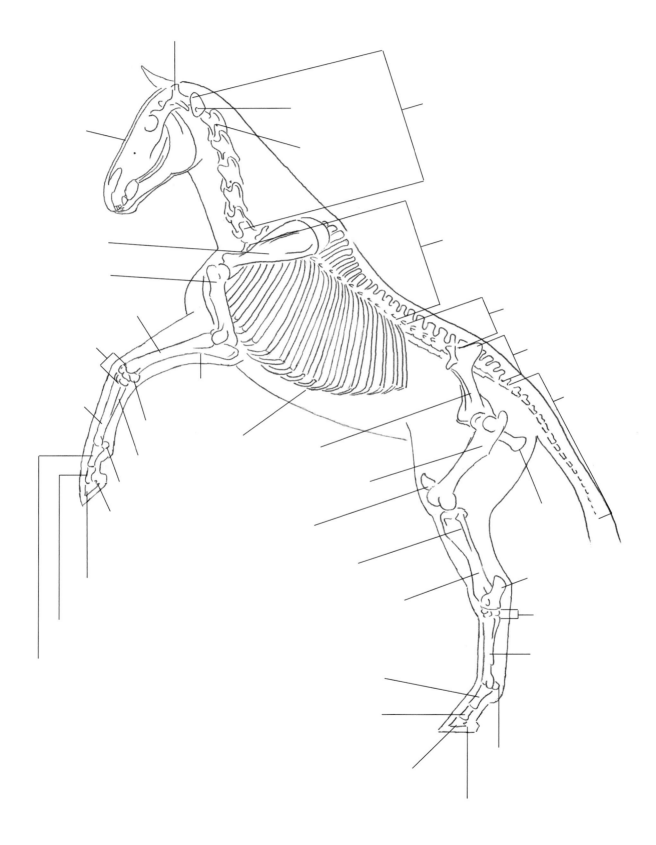

Q8.7 Circle the principles below which, when combined, will help to maintain horse health.

Follow the rules of watering

Exercise correctly

Underfeed the horse – he should look lean

Worm regularly

Clip appropriately

Work the horse on hard ground daily for 1 hour to strengthen the horse's legs

Rug appropriately

Routine checks - dentist, chiropractor

Work the horse in deep going to strengthen the tendons

Regular attention to the horse's feet

Follow the rules of feeding

Q8.8 List the horse's normal temperature, pulse and respiration rates and give some basic ideas about when these may vary.

TEMPERATURE		
PULSE		
RESPIRATION		

Q8.9 Explain why eating, drinking, breathing, droppings and stance can be an indicator of health/ill-health.

BREATHING	
EATING	
DRINKING	
DROPPINGS	
STANCE	

Q8.10 Identify the external parts of the horse's foot, and briefly give a description of the role/function of parts A, C, D, G, H, I and K.

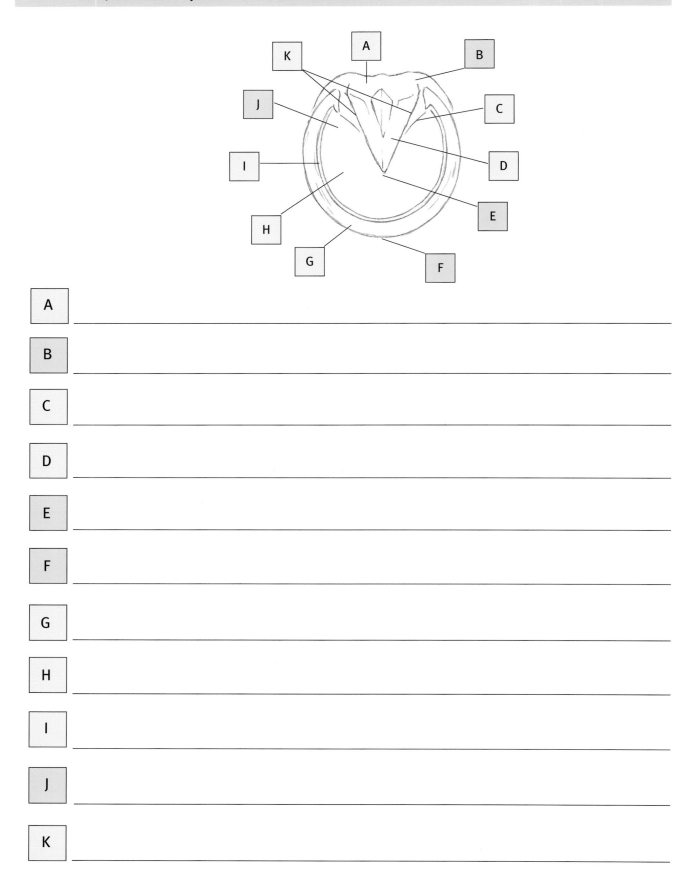

A _____

B _____

C _____

D _____

E _____

F _____

G _____

H _____

I _____

J _____

K _____

9 HORSE BEHAVIOUR

Q9.1 Imagine looking at a group of horses in the field. It is summer. They seem unsettled. What possible reasons could explain this behaviour?

_____

_____

_____

Q9.2 Look at the two images below: one shows a herd enjoying a natural lifestyle, the other shows a horse in a stable. Considering its natural lifestyle, in what ways have we changed how the domesticated horse lives?

	NATURAL LIFESTYLE	STABLED
Exercise programme		
Feed/water		
Breeding		
Grooming		
Contact		
Feet		
Predators/disease		
Instincts		

Q9.3 Write down the ways in which a horse may indicate that he is nervous in the various situations below.

Stable

Field

Ridden in the school

Ridden on the road

Jumping

Being led

Q9.4 List methods of restraint, starting with the mildest.

1. Headcollar	2	3
4	5	6
7	8	9 IV sedation (by vet)

Q9.5 From the answers in Q9.4, list those you would use in the circumstances below, starting with the mildest.

Leading to/from a field

Clipping

Loading

Treatment

Q9.6 A new horse is due to arrive at your yard. What good practices would you follow to accommodate him prior to arrival, on arrival and after one week?

PRIOR KNOWLEDGE

ON ARRIVAL

AFTER 1 WEEK

Q9.7 Why might a stabled horse be difficult to catch when turned out?

_____

_____

_____

_____

Q9.8 List five reasons why a horse may exhibit anti-social behaviour when ridden.

_____

_____

_____

_____

_____

Q9.9 This horse is exhibiting signs that could suggest his tack does not fit or that he is not comfortable. List the tell-tale signs shown and any others you can think of.

_____

_____

_____

_____

_____

_____

_____

_____

_____

10 HORSE HEALTH

Q10.1 Describe the treatment for each of these types of wound.

GALLS	
GRAZES	
BRUISES	
LACERATIONS	
INCISED	
PUNCTURE	

Q10.2 (a) What should you put on a medium lacerated wound while waiting for the vet to arrive?
(b) Which inoculation should you check is up to date?

(a) _____

_____

_____

(b) _____

Q10.3 Look at the picture below and list the rules of basic sick nursing.

_____

_____

_____

_____

_____

_____

_____

_____

_____

_____

_____

Q10.4 Looking at the list of injuries, indicate whether you would call the vet, provide basic treatment or monitor to see how the condition develops.

FIRST SIGNS OF COLIC	
MINOR LAMENESS	
HORSE SEEN QUIDDING	
SEVERE LAMENESS	
MINOR GRAZE TO THE QUARTERS	
SUSPECTED BROKEN LEG	
SERIOUS ARTERIAL BLEED	
HORSE A LITTLE DULL	

Q10.5 What horse records do you consider essential?

_____

_____

_____

_____

_____

_____

_____

_____

_____

Q10.6 Complete the life cycle of the roundworm.

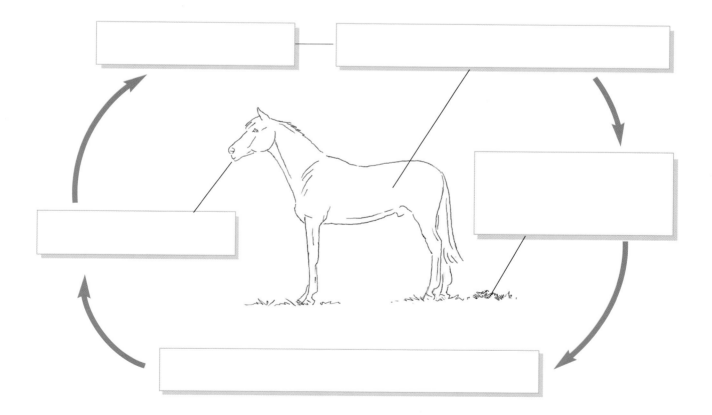

Q10.7 The horse below looks 'wormy'. What are the signs of a worm infestation?

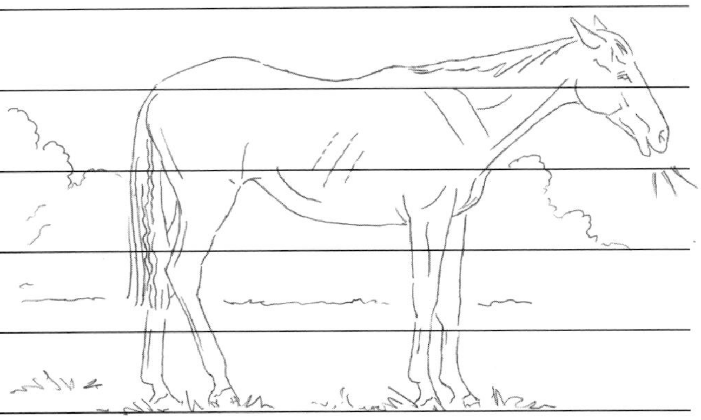

Q10.8 Describe an annual plan for worming a horse: (a) using worm counts; and (b) using base wormer.

(a) _____

_____

_____

_____

(b) _____

_____

_____

_____

_____

Q10.9 What signs might lead you to suspect that a horse's teeth require attention?

_____

_____

_____

_____

_____

_____

Q10.10 What does the dentist/vet do when he/she rasps the teeth and why?

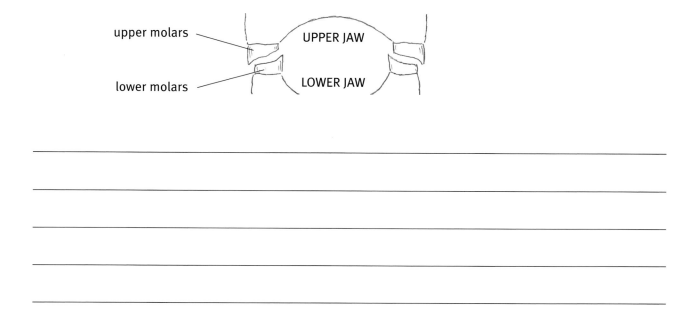

_____

_____

_____

_____

Q11.1 Before starting your fittening programme: (a) what checks would you make?
(b) how would you begin to change the horse's routine?

(a)

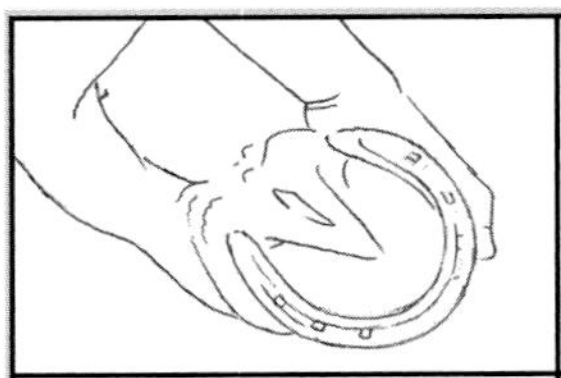	
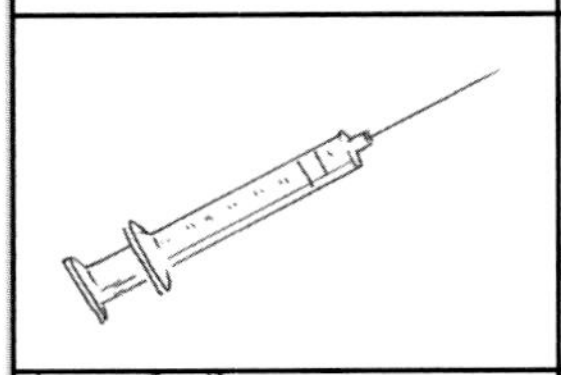	
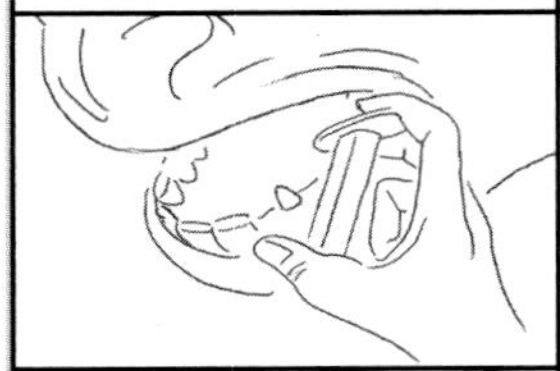	
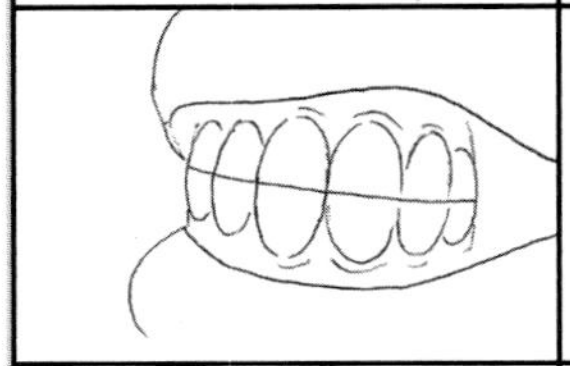	

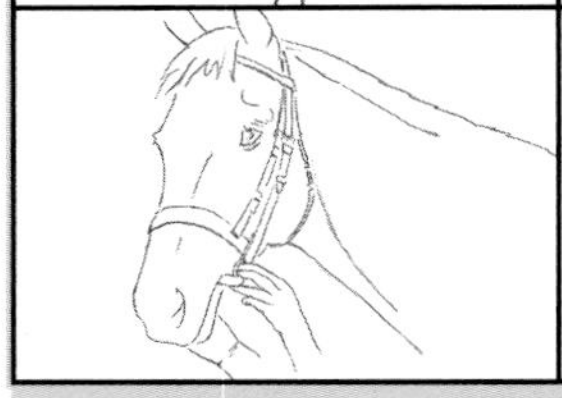	

(b)

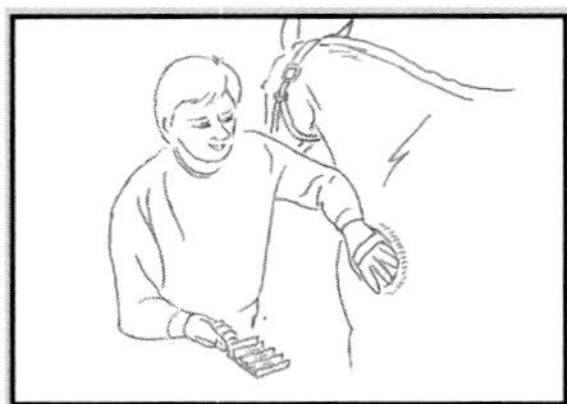		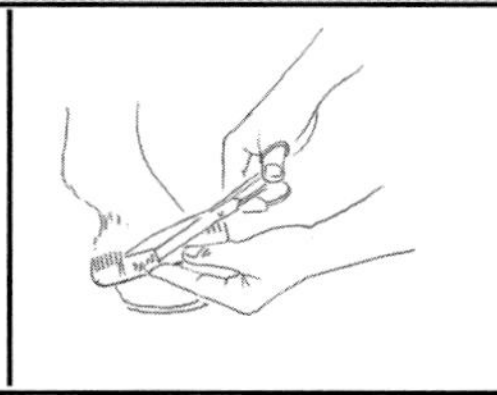

Q11.2 Complete the table to provide an appropriate fittening programme for bringing a horse up from grass into regular work. At the end of the programme, the horse will be capable of being ridden for an hour per day, competing at unaffiliated dressage and show jumping most weekends. (The answer gives just one possible example.)

WEEK	1	2	3	4	5	6	7	8
WALK	15–30 min							
TROT		No						
CANTER	No	No						
LUNGEING	No	No						
SCHOOLING	No	No	No	Light schooling				
HACKING	Yes		Yes					
HILLS							Yes	
JUMPING	No							
FASTER CANTER WORK								

Q11.3 Write out an example plan for weeks 7 and 8 of a fittening programme. (The answer gives just one possible example.)

	MON	TUE	WED	THURS	FRI	SAT	SUN
WEEK 7							
WEEK 8							

Q11.4 Following the 8-week programme in Q11.2, complete the table of feed changes during the programme.

WEEK	Fibre %	Concentrate %	Type of feed
1			
2			
3			
4			
5			
6			
7			
8			

View this as a basic programme which can easily be varied.

Q11.5 **Give four possible reasons why a horse might get coughs/colds when being brought up from grass.**

1. _____

2. _____

3. _____

4. _____

Q11.6 **What are the possible reasons why horses suffer from saddle sores/girth galls when coming up from grass?**

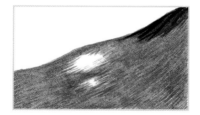 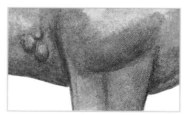

_____

_____

_____

_____

Q11.7 What potential problems caused by dust and/or change of diet might be encountered when first bringing the horse up from grass?

RESPIRATORY	DIGESTIVE

Q11.8 Complete a suitable 3-week roughing-off programme for a fit horse.

WEEK	1	2	3
FEED			
WORK			
RUGS			
SHOES			
TURNOUT			

Q11.9 What daily field checks would you make for a horse that had been roughed off?

_____

_____

_____

_____

_____

Q11.10 Circle the conditions that could possibly cause concussion injuries and draw a rectangle around those that could cause sprain/strain injuries.

Trotting downhill

Fast work in deep going

Deep going

Jumping on a soft surface

Hard ground

Jumping on a hard surface

Fast work on hard ground

Road work in trot

Q11.11 How should a horse be cooled down after routine work?

_____

_____

_____

_____

_____

_____

Q11.12 How should a horse be cared for after his work?

_____

_____

_____

_____

_____

_____

Q12.1 Here is a well-dressed horse and rider, ready to ride on the highway. Label the drawing to highlight the safety aspects.

Q12.2 List the sequence for making a clear manoeuvre on the road.

Q12.3 Describe safe procedure when turning right from a minor to major road.

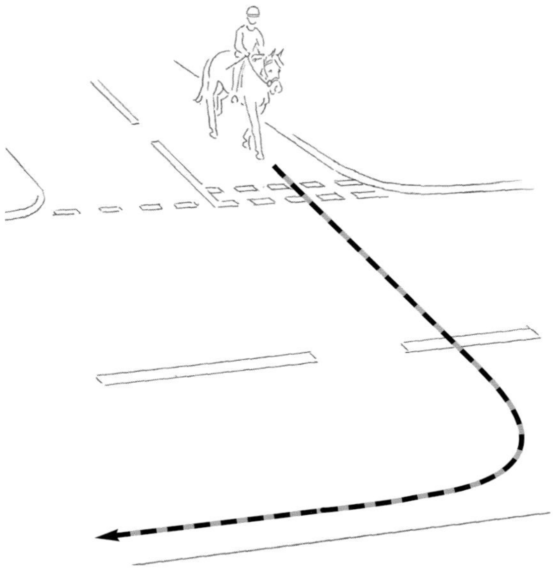

_____

_____

_____

_____

_____

_____

_____

Q12.4 In the same way, describe safe procedure when turning left from a major to minor road.

_____

_____

_____

_____

Q12.5 High visibility clothing – list three items that are available for the rider and five items for the horse.

RIDER _____

HORSE _____

_____

_____

Q12.6 Give two examples of good practice when riding on bridleways.

1. _____

2 _____

Q12.7 List the aims of The British Horse Society.

_____

_____

_____

_____

Q12.8 (a) In the event of an accident you should follow the APACH rules – but what do the letters stand for? (b) Fill in the missing words to complete the paragraph describing a safe procedure in the event of an accident.

(a)

A = _____

P = _____

A = _____

CH = _____

(b)

Airway	safe	accident	aiders	ambulance	Breathing	reacted	pressure
	horses	casualty	Circulation	assessing	nervous	conscious	

When _____ the situation, quickly observe all involved and how they have _____

to the accident. If other riders are _____ they may cause a further _____. To prevent

this it may be necessary to clear the area or request help to hold _____ and reassure riders. Ensure

that the area is _____ before approaching the casualty to assess their welfare. Ascertain if the

casualty is _____. If they are unconscious, send for an _____ immediately, and if

they are seriously bleeding, stem it with _____. Carry out an ABC check – A _____,

B_____, C_____. Ask the conscious _____ how they feel and where

it hurts. Further help will come from the yard's qualified first –_____.

13 GRASSLAND CARE

_____

_____

_____

_____

_____

_____

_____

_____

_____

_____

_____

_____

Q13.2 Complete the table to give a basic yearly plan for maintaining pasture.

WINTER	SPRING	SUMMER	AUTUMN

Q13.3 Why do hedges need cutting?

_____

_____

_____

Q13.4 List routine daily field checks.

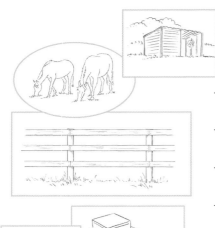

_____

_____

_____

_____

_____

_____

_____

_____

Q13.5 Name each of these common poisonous plants.

_____

_____

_____

_____

Q13.6 Identify these good and bad grasses and tick those that are useful in horse pasture.

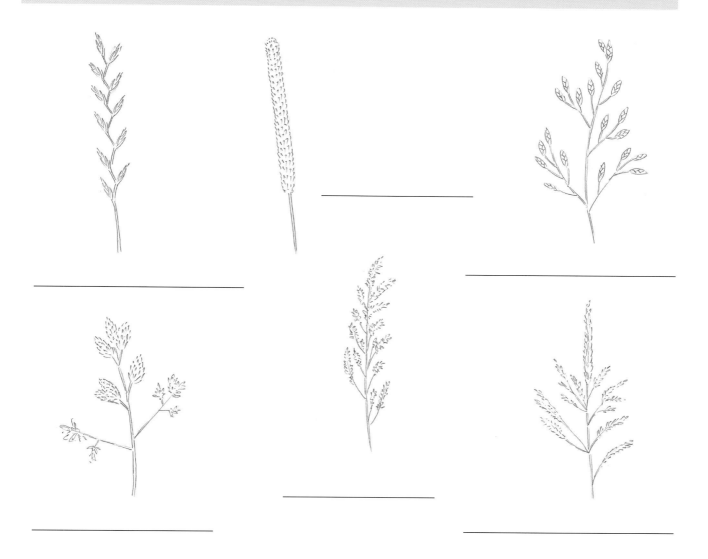

14 WATERING AND FEEDING

Q14.1 List the rules of watering.

_____

_____

_____

_____

_____

_____

_____

Q14.2 The pictures show different methods of watering horses at grass. Write down the advantages and disadvantages of each.

	ADVANTAGES	DISADVANTAGES

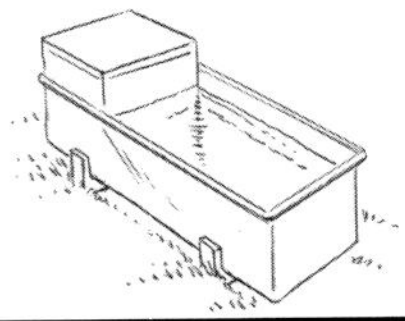		

Q14.3 Complete the table to list the rules of feeding and their reasons.

	RULE	REASON
AMOUNT		
QUALITY		
HYGIENE		
ROUTINE		
CHANGES		
SUCCULENT		
FIBRE		
LITTLE AND OFTEN		
EXERCISE		
WATER		

Q14.4 Fill in the gaps to complete the table of feed values.

FEED	PROTEIN	CARBOHYDRATE	FATS/OIL	HEATING/ FITTENING	FIBRE
OATS	AVERAGE				
BARLEY	AVERAGE	HIGH			
MAIZE				HEATING FATTENING	
BRAN				NO	
SUGAR BEET					
ALFALFA			AVERAGE		
CHAFF					
LINSEED	HIGH		AVERAGE		
SOYA-BEAN MEAL	HIGH				
PEAS AND BEANS					
MOLASSES					

ORM (e.g. flaked, bruised)	DESCRIPTION	MAXIMUM FEED	NOTES
	DARK PELLETS/SHREDS		
		N/A	
			SEEDS NEED COOKING OTHERWISE POISONOUS; LINSEED ADDS SHINE TO HORSE'S COAT
MINCE			

Q14.5 Circle the fibre and concentrates that are traditionally suitable for a horse in hard, fast work.

Seed hay

Meadow hay   Haylage

Barley   Oats   Sugar beet

Stud cubes   Alfalfa   Maize

Peas and beans   Soya-bean meal

Bran

Q14.6 Give the percentage and type of feedstuff that could be fed to two horses at grass in winter. One is on maintenance, one on light work.

	MAINTENANCE	LIGHT WORK
HAY/HAYLAGE		
CONCENTRATE		
TYPES OF CONCENTRATE		

Q14.7 (a) What are the considerations when feeding an older horse?
(b) What fibre and concentrate could be fed to an older horse and why?

(a) _____

_____

_____

_____

(b) _____

_____

_____

_____

Q14.8 What feedstuffs might you feed to a sick horse?

_____

Q14.9 (a) How long should sugar beet pellets be soaked, and in what quantity of water?
(b) How long should sugar beet shreds be soaked, and in what quantity of water?
(c) How long should 'SpeediBeet' be soaked, and in what quantity of water?

(a) PELLETS	(b) SHREDS	(c) SPEEDIBEET
Water	Water	Water
Soak	Soak	Soak

Q14.10 Fill in the blanks to complete the recipe for bran mash and for cooking linseed.

poisonous	split	maintain	water	cooked	limestone flour	water	appealing
six	crumbly	cool	cool	soak	bran	water	apples	phosphorus	water

BRAN MASH

To 2lbs (0.9kg) of _____ , add enough boiling _____ to create a _____ consistency. Add a little

_____to correct the calcium : _____ ratio. Cover and allow to _____until cold enough to

eat. To check consistency, squeeze a handful of cooled bran into a ball. It should _____ the round

form. Add sliced _____ or carrots to make the feed more_____.

LINSEED JELLY

Cover 2–3ozs (55–85g) of dried seed with _____and _____overnight. Add 4 pints (1.8l) of
_____

simmer for _____ until all the seeds _____. Allow to _____ before feeding. To

turn the jelly into a tea, add more _____. Linseed is _____ as dry seed and must be

_____before feeding.

Q14.11 Write captions to the cartoon strip to explain how to soak a haynet.

A _____

B _____

C _____

D _____

Q14.12 Circle the genuine reasons for feeding soaked hay.

Washes spores away so that they are not inhaled

Prevents horses coughing

Swells the hay seeds so that they are digested, not inhaled

Turns it into haylage

Makes poor, dusty hay palatable

Q14.13 The alternative fibres are hay, haylage and silage. State which are suitable for horses and give the different moisture content of each.

	HAY	HAYLAGE	SILAGE
SUITABLE FOR HORSES			
MOISTURE %			

Q14.14 Using the table, make a feed chart for the imaginary scenario below. Use as much or as little of the chart as required.

You have a yard of five horses: Galleon, Luca, Sigi, Dexter and Melody. All are fed three times per day, except Melody who does not have lunch.

Galleon, Luca and Sigi all have 5lbs (2.2kg) haylage am and lunch, and 10lbs (4.5kg) pm. They have 3lbs (1.3kg) of pony nuts am and lunch, and 4lbs (1.8kg) pm.

Dexter has 5lbs (2.2kg) haylage am and lunch, 8lbs (3.6kg) pm. He has three hard feeds, each one 3lbs (1.3kg) competition mix and 1lb (0.45kg) alfalfa. He has 10g of electrolyte supplement am only.

Melody is off work and in the field all day. She is on a maintenance diet of 100% haylage and grass: 4lbs (1.8kg) am, 8lbs (3.6kg) pm. To stop her banging she has two carrots when the others get fed.

	GALLEON		LUCA		SIGI		DEXTER		MELODY	
	am + lunch	pm	am + lunch	pm	am + lunch	pm	am + lunch	pm	am	pm
HAYLAGE										
PONY NUTS										
COMPETITION NUTS										
ALFALFA										
SUPPLEMENTS										
CARROTS										

15 RIDING

Q15.1 Describe how to quit and cross your stirrups.

_____

_____

_____

_____

_____

_____

_____

Q15.2 List occasions when you may need to ride with the reins in one hand.

_____

_____

Q15.3 The statements below have been made about the use of the whip. Write TRUE or FALSE at the end of each statement.

The whip is an object to inflict pain on the horse

The whip is used only to make a horse go faster

The whip should be used behind the leg to reinforce the leg aid

The use of the whip should not interfere with the contact on the reins

The whip should be changed over during the change of rein

Q15.4 Circle the words you would pick to describe the qualities required to gain a horse's trust when riding.

Routine

Being quick tempered

Negative reinforcement

Honesty

Repetition

Consistency

Positive reinforcement

Patience

Indecision

Q15.5 There are rules for riding in open order. List as many as you can.

_____

_____

_____

_____

_____

Q15.6 The picture shows a rider adopting a balanced position (also known as forward seat, jumping position). Annotate each label to describe how each area leads to a correct position.

HEAD

SHOULDERS

BACK

HIPS

KNEES

LOWER LEG

ANKLES

HEELS

Q15.7 List the rules for riding in the open in a group.

_____

_____

_____

_____

Q15.8 Undulation and ground conditions can influence the horse's speed and balance. Using a line, link the ground to the correct description.

DOWNHILL	Unbalanced
UPHILL	Faster, less balanced
SLIPPERY, GREASY	Slower, balanced
HEAVY, DEEP	Slower or faster, balanced
HARD	Slower, unbalanced

Q15.9 Why should you shorten your stirrups when preparing to jump?

_____

_____

Q15.10 Fill in the gaps to complete the paragraph, describing how to approach, jump and depart from a fence.

The canter should be forward, balanced and _____. During the turn on the approach to the fence,

the_____ of the canter is maintained by keeping the horse balanced between leg and hand. On the

straight line approach, again the quality of the _____ is maintained. The rider guides the horse to the

_____ of the fence and should _____ _____ and over the jump, focusing on the departure. As the

horse _____ , the rider moves forward into _____ position, allowing the horse's _____

to move forward whilst maintaining a light _____ at the end of the reins. The rider helps the horse to

jump by remaining in _____ with the horse. Maintaining a _____ lower leg position allows the

rider to land in _____ and immediately rebalance the horse between leg and hand, recognise the

canter _____ and change it if necessary, using the next _____. The canter rhythm is re-

established, if necessary, and a straight line ridden _____ from the fence, maintaining the quality of the

_____ through the departure turn.

Q15.11 What might be the result of a rider communicating a lack of commitment or apprehension on the approach to a fence?

_____

_____

_____

_____

_____

Q15.12 What factors do you believe lead to a fluent show-jumping round?

RIDER

_____

_____

_____

_____

HORSE

_____

_____

_____

_____

_____

1 GROOMING

Q1.1

	QUARTERING	STRAPPING	BRUSHING OFF
HEAD	YES	YES	YES
BODY	YES	YES	YES
MANE AND TAIL	YES	YES	YES
LEGS	YES	YES	YES
FEET PICKED OUT	YES	YES	YES
FEET WASHED AND OILED	NO	When necessary	NO
EYES AND NOSE SPONGED	YES	YES	YES
DOCK SPONGED	YES	YES	YES
STABLE STAINS SPONGED	YES	YES	If bad
STABLE RUBBER	NO	YES	NO
BANGING	NO	YES	NO
TIME TAKEN	10 min	30 min–1 hr	5 min

Q1.2

1 Tie up the horse.
2 Keep grooming kit tidy and together.
3 Work on a non-slip floor.
4 Move the horse around so that you never work in a confined space.
5 Wear correct footwear.

Q1.3

The equestrian industry wants to employ efficient staff – horses are labour intensive so staff need to work within time scales.

Q1.4

TACK	Loosen the girth, run up the stirrups and undo the noseband.
WALK	Walk the horse until he has stopped blowing (approximately 10 mins).
WATER	Offer the horse a third of a bucket of chilled water when he has stopped blowing.
UNTACK AND WASH	Remove all tack and wash down with tepid water, checking for cuts and swellings. Sweat-scrape excess water.
RUG	Rug appropriately for the time of year, using a cooler.
WATER	Offer half a bucket of water on return to stable.
TACK	Clean now if possible. It will be harder to clean once the sweat has dried.
INSPECT	Return to the horse, re-inspect and check temperature. Report if anything found.
WATER	Ad lib water now.
GROOM AND RUG	Groom and re-rug as necessary when dry.

Q1.5

(a)

We use an odd number so that the eye does not visually cut the neck in half.

(b)

Long neck – fewer plaits – looks shorter.
Normal neck – 9/11/13 plaits.
Short neck – more plaits – looks longer.

(c)

A Divide the mane into equal sections.
B Plait down and secure with a band/thread.
C Fold in half.
D Fold in half again.
E Secure with a band/thread.

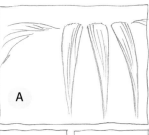
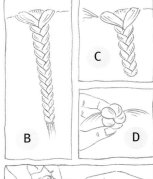

Q1.6

(a)

In a 'flat' plait the new sections are taken over and into the plait. In a 'ridge' plait the new sections are taken under and into the plait.

(b)

The first is straight, with correct-length equal sections used, tight all the way down. The second is crooked, too short, loose, and plaited using different widths of hair.

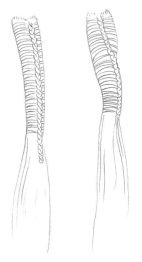

2 CLOTHING

Q2.1

LIGHTWEIGHT NZ	Spring/summer/autumn rain sheet; also keeps horse clean.
MIDDLEWEIGHT NZ	Winter. Stabled horse turned out for a few hours per day. Not fully clipped. Native ponies with full coat turned out in the winter.
HEAVYWEIGHT NZ	Winter. Stabled, fully clipped and turned out a few hours in the day. Neck and belly clip living out in winter.
SUMMER SHEET	Summer. Stabled. Keeps flies and dust off.
LIGHTWEIGHT COOLER	After work/bath in spring and autumn. Stabled.
HEAVYWEIGHT COOLER	After work in winter. Stabled.
LIGHTWEIGHT STABLE RUG	Stabled spring/autumn.
HEAVYWEIGHT STABLE RUG	Stabled winter.
STABLE NECK COVER	Stabled winter.

Q2.2

1 Tie the horse up.
2 Fold the blanket into quarters and position over the withers, with the top of the blanket at the base of the horse's ears.
3 Unfold over the horse's back.
4 Fold the front corners up to the withers.
5 Put on top rug and fasten.
6 Fold blanket triangle back over the withers.
7 Fasten surcingle over the folded triangle and girth to hold the blanket in place, with a wither pad to protect the horse's back.

Q2.3

Protection. Drying out. Reducing filling. Warmth.

Q2.4

A Wrap Fybagee around the leg and fit. Overlap in the direction of bandaging, i.e. front over back.
B Begin first wrap of bandage with the end left up and out.
C Fold the protruding flap over.
D Continue to bandage, overlapping approximately half a bandage each time.
E As you reach under the fetlock, make a 'V' with the bandage at the front of the pastern.
F Fasten and finish.
G Check that there is the same firm pressure at the top and bottom.

Q2.5

Fybagee for travelling can be shaped to fit over the knees and hocks, whereas Fybagee for stable bandages would not be shaped.

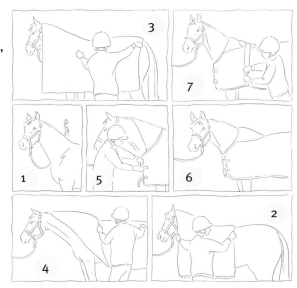

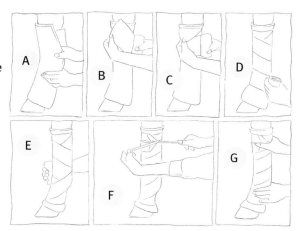

Q2.6

POLL GUARD – Prevents the horse from hurting his head if he should throw his head up and hit the ceiling of the horsebox/trailer.
LEATHER HEADCOLLAR – Leather breaks in an emergency.
RUG – Appropriate for time of year; maintains body temperature.
ROLLER (with wither pad) – Holds tail guard in place.
TAIL GUARD – Gives extra protection to tail.
TAIL BANDAGE – Protects the tail from rubbing.
TRAVEL BOOTS/BANDAGES – Protect legs/hocks/knees/fetlocks/coronets from knocks.

Q2.7

Tie the horse up securely. On clean, dry legs fit each boot individually, positioning correctly and then fastening. When fastening, start with the middle strap and work up and down the leg from the centre out. Fasten firmly into position, ensuring no slippage. Bandages are often considered safer than boots.

Q2.8

The horse may tread on the boots and lose his footing.
The horse may need to alter his position and find that he cannot as he is treading on the opposite boot.
Poor quality boots may not stay up, leading to parts of the legs being unprotected.

Q2.9

RUGS – Oil leather parts and grease buckles. Machine wash; for NZs use a specialist company.
BOOTS – Synthetic: machine wash and air dry; leather: wash and saddle soap.
BANDAGES – Machine wash and air dry.
FYBAGEE – Machine wash and air dry.
ALL should be stored in a cool, dry, vermin-proof area with moth balls. Leather oiled intermittently.

3 SADDLERY

Q3.1

- SIZE – Must be the right size for the horse.
- LEVEL – Pommel and cantle level.
- POMMEL CLEARANCE – Four fingers' width between pommel and withers.
- GULLET – Clear daylight through gullet.
- SHOULDER – Fits around shoulder.
- BEARING SURFACE – Even bearing surface from behind.

Q3.2

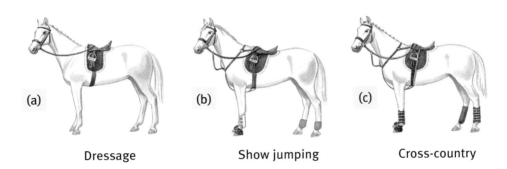

(a)   (b)   (c)

Dressage   Show jumping   Cross-country

Q3.3

Horse is unhappy and does not perform well.
Horse's attitude changes – he becomes grumpy.
Horse bucks when ridden.
Sores develop, e.g. saddle, girth galls, and sores under the bridle.
Muscle wastage over the topline because the horse works incorrectly.

Q3.4

BRIDLE – Fitted as normal, with a snaffle bit. Noseband removed. Reins tied up in the throatlash.
CAVESSON – Noseband fitted two fingers' width below the projecting cheek bone. Buckles fastened under the cheek pieces and under the jaw, firmly.
SADDLE – Fitted as normal, stirrups tied up.
SIDE REINS – Attached under first girth strap and around the third. Approximate length 2ins (5cm) short of the base of the ear for elastic side reins. Clipped to the 'D' ring, across withers, when not in use.
BOOTS – Brushing, fitted as normal.
LUNGE LINE – Usually clipped onto centre ring of cavesson.

Q3.5

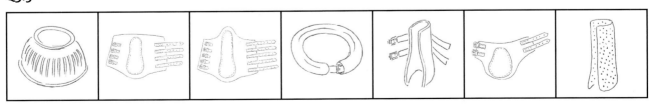

Over-reach   Brushing   Speedicutting   Sausage   Tendon   Fetlock   Porter

OVER-REACH – Protect the heel from over-reaching injuries.
BRUSHING – Prevent damage when the horse brushes. Some have a reinforced plate along the front of the hind leg, for eventing, and down the back of the front leg.
SPEEDICUTTING – High boots that protect the area just below the hock - the site for speedicutting injuries.
SAUSAGE – Put on one leg only, to help stop injuries to the coronet from the opposite foot.
TENDON – Used for show jumping. Protect the tendons down the back of the leg, whilst allowing the horse to feel if he knocks a pole with his cannons.
FETLOCK – For hind legs only; shortened boots that cover the fetlock only. Used solely for show jumping.
PORTER – Synthetic eventing boot, light-weight and durable, held in place by a bandage.

Q3.6

RUNNING MARTINGALE
• Neckstrap joins martingale in the centre of the chest, where neck meets chest.
• Four fingers' width between neckstrap and wither.
• Martingale ring to reach between crest and hand span width below.

STANDING MARTINGALE
• Neckstrap joins martingale in the centre of the chest, where neck meets chest.
• Four fingers' width between neckstrap and wither.
• Length of martingale – not less than when measured by running it under the throat and down the jaw to the noseband.

Q3.7

• Top of breastplate should sit just in front of the sensitive wither area.
• One hand's width between the neck strap and the horse.
• Girth loop should be approximately a hand's width from the chest.
• Martingale attachment should come up vertically either side to a minimum of one hand's breadth below the crest.

Q3.8

WARM WATER – Cleans the tack, removes sweat and grime.
OIL – Supples and waterproofs.
SADDLE SOAP – Supples, waterproofs and shines.

Q3.9

Tack cleaned and heavily oiled.
Store under cover to prevent accumulation of dust, in a clean, dry area that will maintain a constant temperature.

4 HANDLING AND LUNGEING

Q4.1

	MUCKING OUT	GROOMING	LEADING
RISK			
SELF	Horse could tread on you. Be aware. Always work in plenty of space or tie the horse outside the stable.	Horse could step on or kick you. Tie up. Work in plenty of space.	External influence on the horse. Look for hazards. Anticipate.
HORSE	Tools. Use with care.	Horse could step on kit. Leave grooming kit outside.	Hazards when leading. Change route. Remove hazards.
OTHERS	Tools used with care. Put away afterwards.	Someone could step on kit. Put kit away after use.	Horse reacts unexpectedly. Move horses at times when minimal number of people around.

Q4.2

Patient. Quiet. Relaxed. Efficient. Organised. Assertive.

Q4.3

To observe the horse working.
The horse has a sore back/saddle sores.
To improve suppleness.
To develop responsiveness to the voice.
When breaking in.
To add variety in the work.
Rider unable to ride.
To take the edge off a fresh horse.
Build up muscle.
Discipline.

Q4.4

A	B
Standing tall	Slouching
Eye contact	Eyes to the ground
Hands in correct position	Hands too high
Shoulders back, looking assertive	Round shoulders, meek
Whip in position to be effective	Whip ineffective
Standing square to the horse	Standing side on

Q4.5

Consistent. Accepting.

Q4.6

The horse must stay at the end of the lunge line.
The horse must respond correctly to commands.

Q4.7

WALK – Warm up and cool down. The horse must get used to walking on the lunge.
TROT – Exercise, suppleness, balance and rhythm. The main pace used when lungeing.
CIRCLE – Suppleness, rhythm, balance, engagement.
TRANSITIONS – Responsiveness to the aids, engagement.
CANTER – Can be used to encourage more energy in the trot in a lazy horse, or to take the edge off a fresh horse.

Q4.8

CAVESSON – Leaves the bit free for the side reins. No pressure on the mouth.
SIDE REINS – Provide the horse with a contact to work into and therefore produce an outline and develop a correct way of going. Acceptance of the contact.
BRUSHING BOOTS – Protection.
LUNGE LINE – Dictates the shape and size of the circle.
SADDLE/ROLLER – Introduction to tack. Point of attachment for the side reins. Saddle required if lungeing a rider.
LUNGE WHIP – Reinforces the lunger's voice.

Q4.9

Protection.

5 STABLE DESIGN

Q5.1

	ADVANTAGE	DISADVANTAGE
WALLS		
Breeze block	Relatively inexpensive	Porous, therefore not suitable for external walls
Brick	Excellent temperature regulation; durable	Expensive
Wood	Cheap, quick to erect	Not fire resistant; poor temperature regulation; encourages wood chewing
ROOF		
Tiles	Good temperature regulation	Potential hazard in high wind
Plywood and felt	Relatively inexpensive	Fire hazard
Corrugated iron	Inexpensive; allows light filtration when interspersed with clear plastic sheeting	Noisy in rain; does not aid temperature regulation
FLOOR		
Brick	Durable; fireproof; good temperature regulation	Expensive; not ideal for horses to stand on for long periods
Concrete	Durable; less expensive than bricks	Not ideal for horses to stand on for long periods
Dirt	Inexpensive; good drainage therefore less bedding used	Possibly muddy if allowed to become wet

Q5.2

	16 hh HORSE	PONY
SIZE	12ft x 12ft (3.6 x 3.6m)	10ft x 12ft (3m x 3.6m)
HEIGHT	15ft (4.5m)	12ft (3.6m)
DOOR WIDTH	4ft (1.2m)	4ft (1.2m)
DOOR HEIGHT	10ft (3m)	8ft (2.4m)
HALF DOOR HEIGHT	4ft 6ins (1.3m)	3ft–3ft 6ins (0.9m–1.06m)

Q5.3
(a)
Half door.
Slatted window and bars.
Ventilated corrugated sheeting.
Roof vents.
Louvre boards.

(b)
The horse should not breathe stale air.
Prevents the spread of disease and aids recuperation.

Q5.4

(a)

To remove waste water. Slopes backward so urine is removed and horse does not stand in it when looking over the door.

(b)

Stagnant waste water.

Ill-health and disease.

Damp environment.

Respiratory disorders.

Feed/hay becomes damp and turns sour.

Use more bedding.

Q5.5

ESSENTIAL	EXTRA
Tie ring with string	Extra tie ring for haynet
Watering system – buckets/automatic waterer	Feed manger
	Hay rack

Q5.6

Less to harm the horse.

6 SHOEING

ANVIL

FIRE TONGS

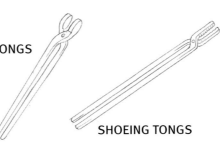
SHOEING TONGS

BUFFER

DRIVING HAMMER

PINCERS

Q6.1

A ANVIL – Iron block upon which the shoes are shaped.

B FIRE TONGS – Used to move the shoes in and out of the forge.

C SHOEING TONGS – Hold the shoe when shaped on the anvil.

D BUFFER – Raises clenches.

E DRIVING HAMMER – Used with buffer to raise clenches. Hammers nails into hoof. Trims the ends of nails.

F PINCERS – Levers off the shoe.

G HOOF CUTTERS – Trim the hoof. (One side has a sharp blade; the other has a square end acting as a block.)

H DRAWING KNIFE – Trims the horn, sole and frog.

I PRITCHEL – Carries the hot shoe.

J RASP – Levels the surface of the foot and finishes off around the edge once the shoe is on.

K NAIL CLENCHER – Folds the nail ends over so the clench is flush with the hoof wall.

HOOF CUTTERS

DRAWING KNIFE

PRITCHEL

RASP

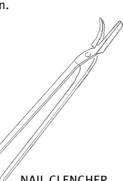
NAIL CLENCHER

Q6.2

A Foot assessed by the farrier.

B Clenches raised using buffer and driving hammer.

C Shoe removed using pincers.

D Shoe put into the forge using fire tongs.

E Excess growth is trimmed using the hoof cutters.

F Horn, frog and sole trimmed using the drawing knife and an indent is made for the quarter or toe clips.

G The foot is rasped level.

H The hot shoe is shaped as necessary.

I The fit checked by burning into the hoof.

J The cooled shoe is nailed on, and the excess nail cut off.

K The foot is clenched up using the nail clencher (or hammer and closed pincers).

L The foot is finished off with the rasp.

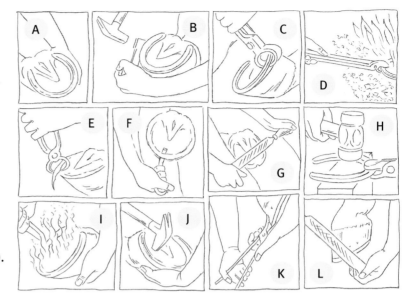

Q6.3

Place foot between knees if front foot, rest on thigh if back foot.

Raise all clenches.

Use the pincers to remove the shoe, levering both heels a quarter of the way down the shoe, inwards.

Q6.4

| Long toe | Overgrown shoe | Risen clenches | Collapsed and contracted heels | Loose shoe |

Q6.5

LONG TOE – Pressure on the tendons and ligaments at the back of the leg and on the hoof wall. Broken hoof/pastern axis.

OVERGROWN SHOE – The shoe sits inside the wall, on the sole, causing bruising.

RISEN CLENCHES – The shoe becomes loose and the movement can cause bruising.

COLLAPSED AND CONTRACTED HEELS – Absorption of concussion, grip and circulation are reduced, which can lead to navicular.

LOOSE SHOE – Shoes come off, often breaking away some of the hoof. Can be dangerous to ride, as the horse will lose purchase on the ground. The horse could go lame if he were to stand on a sharp object.

Q6.6

Foot fits shoe. FALSE (most of the time, unless remedial)

Clenches should be flush with the foot. TRUE

The bearing surface is level. TRUE

The farrier has chosen the correct type and weight of shoe for the horse. TRUE

The horse is sound. TRUE

The shoe fits the foot. TRUE

The front feet look oval, the hind, round. FALSE

Q6.7

Grass crack Chipped/split hoof Long toe Collapsed and contracted heels Foot/pastern axis is broken

Q6.8

IRON – Durable and malleable when hot.

CONCAVE INNER EDGE – Sloping to match the sole, therefore less suction when in mud.

FULLERED – Aids purchase and lightens the shoe.

TOE CLIPS – Hold the shoe in position.

QUARTER CLIPS – Usually hold the hind shoe in place, but reduce the amount of damage inflicted if the horse over-reaches.

7 CLIPPING AND TRIMMING

Q7.1

CORRECT
Shows off the horse's neck.
Thins the mane.
Makes the mane lie better.
Easier to plait.

INCORRECT
Makes the mane longer.
Makes the mane thicker.
Makes the mane less greasy.

Q7.2

Some native breeds. (Native breeds are not shown trimmed.)
In winter a horse who lives out. The hair provides protection and warmth.

Q7.3 & 7.4

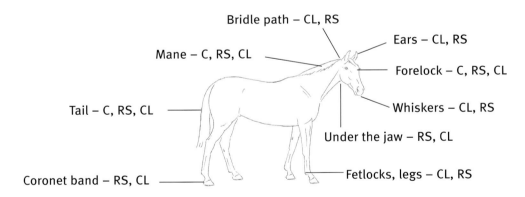

Bridle path – CL, RS
Ears – CL, RS
Mane – C, RS, CL
Forelock – C, RS, CL
Whiskers – CL, RS
Tail – C, RS, CL
Under the jaw – RS, CL
Fetlocks, legs – CL, RS
Coronet band – RS, CL

Q7.5

SWEAT/CHILL – Remove a partial amount of the coat, appropriate for work and lifestyle and reduce the amount sweated. Therefore the horse is less likely to chill.

AESTHETIC – For similar clips, for example the chaser and trace, the choice can improve the appearance of the horse. The trace was designed to follow the lines of the traces for driven horses.

MEDICAL REASONS – Treatment of skin condition or to have closer access to internal structures when scanning.

CLEANLINESS – Easier to keep the horse clean and to groom him.

Q7.6

FULL – All off.
HUNTER – Legs left on. Optional saddle area.

Q7.7

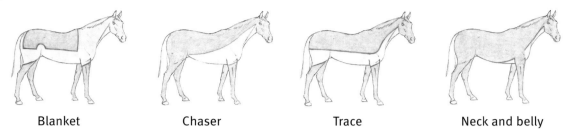

Blanket   Chaser   Trace   Neck and belly

Q7.8

CLIP	LIVE IN/OUT	AMOUNT OF WORK
BLANKET	Live in. Turn out.	Medium.
CHASER	Live in. Turn out.	Medium.
TRACE	Partial in/can be out overnight in good weather.	Light/medium.
NECK AND BELLY	Live out/majority out.	Light.

Q7.9

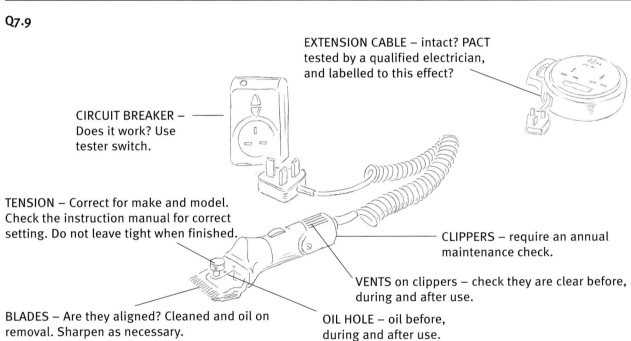

EXTENSION CABLE – intact? PACT tested by a qualified electrician, and labelled to this effect?

CIRCUIT BREAKER – Does it work? Use tester switch.

TENSION – Correct for make and model. Check the instruction manual for correct setting. Do not leave tight when finished.

CLIPPERS – require an annual maintenance check.

VENTS on clippers – check they are clear before, during and after use.

BLADES – Are they aligned? Cleaned and oil on removal. Sharpen as necessary.

OIL HOLE – oil before, during and after use.

Q7.10

YOU	HORSE	HANDLER
Dust mask	Clean and dry	Always stand same side as person clipping
Long hair tied back	Non-slip floor	Pay attention to what is going on
Safe footwear	Tail bandaged	Safe footwear
Overalls – protect clothing	Mane plaited	Dust mask
Move the horse around the working area Never work in a confined space		Watch the horse's reactions all the time

8 ANATOMY AND PHYSIOLOGY

Q8.1

Bright, alert demeanour.
Shiny coat.
No discharge from the eyes or nose.
Even weight-bearing over all four feet. (Resting a hind foot may be normal.)
Pink mucous membranes.
Normal, regular breathing.
Normal drinking.
Normal droppings – number and consistency.
Normal urine – amount and colour.
Normal bed.
With the herd, grazing (if in field).

Q8.2

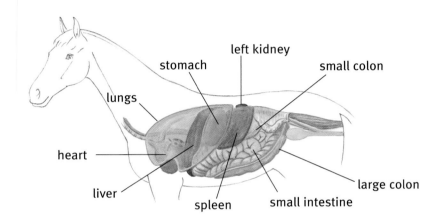

Q8.3

A LIPS – Gather food.
B INCISORS – Crop grass.
C SALIVARY GLANDS – Lubricate and start to break down food.
D MOLARS – Grind.
E TONGUE – Passes bolus of food to the back of the mouth.
F EPIGLOTTIS – Covers trachea as food passes over and into the oesophagus.
G OESOPHAGUS – Via peristalsis (waves of muscular contractions along the oesophagus), passes food from the mouth to the stomach.
H CARDIAC SPHINCTER MUSCLE – One-way valve at entry to stomach, preventing the horse from being sick.
I STOMACH – Gastric juices begin the breakdown of food, approximately 30 min.
J PYLORIC SPHINCTER – Valve between stomach and small intestine.
K SMALL INTESTINE – Further food breakdown by enzyme secretions; absorbs carbohydrates, proteins and fats.
   – Duodenum – Bile (from liver) breaks down fats and oils.
   – Jejunum – Absorbs amino acids, glucose, vitamins and minerals. Villi offer greater surface area for absorption.
   – Ileum – Continues breakdown and absorption.
L ILEOCAECAL VALVE – between small and large intestine.

Q8.4

A CAECUM – Bacteria break down cellulose in grass, hay and haylage – hence need to change diet slowly as bacteria need to change. Nutrients and water absorbed.
B LARGE COLON – Further bacterial breakdown, nutrients and water absorbed. Pelvic flexure is a narrowing in the colon and possible area of blockage.
C SMALL COLON – Absorbs electrolytes and water.
D RECTUM – Waste matter stored before expulsion.
E ANUS – Sphincter muscle regulates waste expulsion.

Q8.5
It slows the digestive system, allowing all food to be broken down and absorbed efficiently. The digestive system is unable to function correctly without bulk.

Q8.6

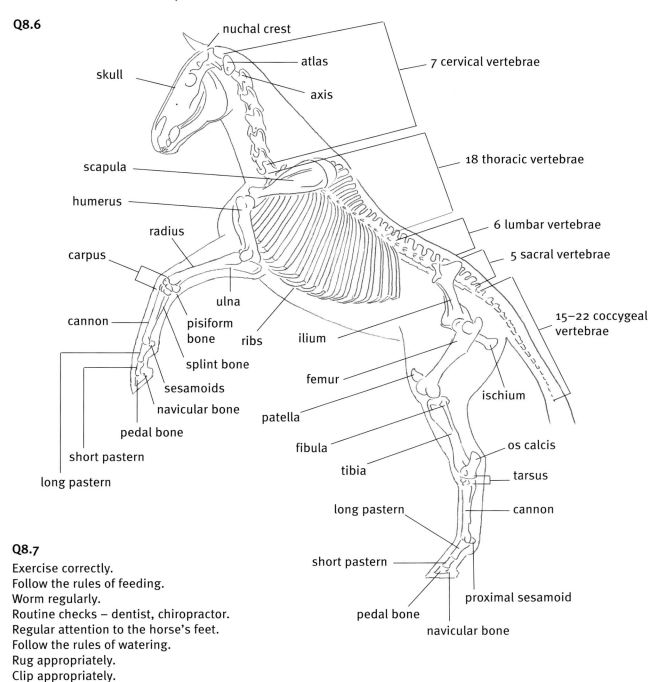

nuchal crest
atlas
7 cervical vertebrae
skull
axis
18 thoracic vertebrae
scapula
humerus
6 lumbar vertebrae
radius
5 sacral vertebrae
carpus
15–22 coccygeal vertebrae
cannon
ulna
pisiform bone
ribs
ilium
splint bone
femur
sesamoids
navicular bone
ischium
pedal bone
patella
short pastern
fibula
os calcis
long pastern
tibia
tarsus
long pastern
cannon
short pastern
proximal sesamoid
pedal bone
navicular bone

Q8.7
Exercise correctly.
Follow the rules of feeding.
Worm regularly.
Routine checks – dentist, chiropractor.
Regular attention to the horse's feet.
Follow the rules of watering.
Rug appropriately.
Clip appropriately.

Q8.8

TEMPERATURE	38°C 100.5°F	Time of day, raised when worked, raised if fever, higher in foals
PULSE	36 – 42 BPM	Time of day, raised when worked, raised if ill, anxious, higher in foals, lower if very fit
RESPIRATION	8 –15 BPM	Time of day, raised when worked, raised if ill, anxious, higher in foals, lower if very fit

Q8.9

BREATHING	Increases due to pain/respiratory problems
EATING	An unwell horse does not want to eat normally
DRINKING	Increase/decrease in amount drunk could indicate ill-health
DROPPINGS	More/less than normal or change in consistency suggests a change in diet or digestive system
STANCE	Resting a hind leg is normal, but not the same one all the time. Resting a foreleg ('pointing a foot') indicates a problem. Head down and dull or 'tucked up' indicates ill-health

Q8.10

A HEEL – Absorbs concussion.
B BULB OF HEEL.
C BAR – Weight-bearing area and provides purchase on the ground.
D FROG – Spongy, triangular cushion; provides grip, absorbs concussion and pumps fluid back up the leg when horse moves.
E POINT OF FROG.
F TOE.
G WALL OF HOOF – Weight bearing. Protects internal structures.
H SOLE – Concave, to allow for the spread of the foot as it comes in contact with the ground; protects internal, sensitive areas.
I WHITE LINE – Divides sensitive and insensitive parts of the foot.
J SEAT OF CORN.
K CLEFT – Aids purchase on the ground.

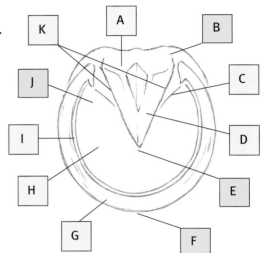

9 HORSE BEHAVIOUR

Q9.1

Flies. Heat. Bickering.

Q9.2

	NATURAL LIFESTYLE	STABLE
Exercise programme	Nomadic	Structured exercise
Feed/water	Nomadic/grazers	Eats when fed
Breeding	Natural selection	Breeding programmes
Grooming	Groom each other	Grooming from staff, including trimming as required
Contact	Within herd	Only with companions during turnout
Feet	Natural wear	Artificial shoeing
Predators/disease	Predators in the wild/survival of the fittest	No predators/veterinary treatments
Instincts	Follow natural instincts	Overcome natural instincts through training, e.g. clipping, wearing rugs

Q9.3

STABLE – Rolls eyes. Head high. Body tense. Turns quarters in fear, ready to fight because he cannot flee.
FIELD – Runs to the furthest end of the field away from danger, then turns and looks.
RIDDEN IN THE SCHOOL – Head up, tries to flee.
RIDDEN ON THE ROAD – Spooks. Bolts.
JUMPING – Refuses, runs out, jumps higher than necessary.
BEING LED – Head up, tries to flee, spooks.

Q9.4

1 Headcollar	2 Halter	3 Bridle
4 Chifney	5 Lift a leg	6 Twitch neck by taking a pinch of skin
7 Twitch nose	8 Sedate orally (by vet)	9 IV sedation (by vet)

Q9.5

LEADING TO/FROM A FIELD – Headcollar, halter, bridle, chifney.
CLIPPING – Headcollar, halter, bridle, lift a leg, twitch neck, twitch nose, call vet – oral sedation, IV sedation.
LOADING – Headcollar, halter, bridle, chifney.
TREATMENT – As clipping.

Q9.6

PRIOR KNOWLEDGE:
Find out which bedding, hay and feed he is used to. Order in advance. Any vices? Temperament?

ON ARRIVAL:
Before unloaded, check passport and vaccination records.
Unload. Check horse over. Trot up to check sound.
Put in isolation stable for two weeks and follow isolation procedures.
Worm according to weight as required.
Ask owner to fill in form listing:
      All equipment brought with the horse
      Dates of last worming, vaccinations, dentist, farrier, chiropractor
      Turn out requirements
      Any ongoing treatments
      Any problems – anti-social behaviour, led in chifney, etc.

AFTER 1 WEEK:
Take temperature, pulse and respiration as normal for comparison.

Q9.7

Excited and enjoying being in the field. Likes companionship. Does not want to return to the stable. Frightened of the person catching him.

Q9.8

Not socialised with other horses.
Not socialised in a ridden environment.
Intimidated by other horses coming too close.
Views the ride as a herd and tries to establish a pecking order.
Uncomfortable with tack/rider and is therefore grumpy.

Q9.9

Nips as the girth is done up from the ground.
Bucks.
Rears.
Bolts.
Open mouth.
Humps over back.
Takes short strides.
Not wishing to go forwards.
Reluctant to respond to aids.

10 HORSE HEALTH

Q10.1

GALLS	Saline wash, antibacterial cream. If infected, call vet for stronger treatment. Do not work with a girth until healed. Ensure in future girth is clean and use girth sleeve if sensitive.
GRAZES	Usually have foreign particles around wound. Cover wound and wash around it, then flush wound with running water to remove foreign particles. Antibacterial cream/spray once clean and dry if superficial.
BRUISES	Cold hose to reduce inflammation. If in an area to which you can apply a cold compress then apply one 20min per hour for the first few hours. Never apply frozen pack directly to skin as it can burn. Use clean gauze between leg and pack. Once the pain is eased, horse will benefit from walking and/or massage to disperse filling.
LACERATIONS	If stitching required, call vet. Otherwise, clean with saline solution, trickle hose. Sterile dressing as appropriate. Monitor, as high risk of infection.
INCISED	Assess if stitching needed and call vet if necessary. If not, treat as lacerations.
PUNCTURE	If near joint or deep, call vet. Otherwise, hot poultice to heal from inside out. Monitor, as high risk of infection.

Q10.2

(a) Nothing, unless you need to stop bleeding or it could become further infected without a dressing, in which case, use a sterile, dry dressing and apply pressure as necessary. Do not use creams or coloured sprays otherwise the vet cannot see the wound clearly.

(b) Tetanus.

Q10.3

Convalescence diet: 100% fibre immediately. Roughage – meadow hay – easy to chew and palatable.
Constant supply of fresh, clean water.
Clean, hygienic environment.
Ventilation without draughts.
Keep the horse warm – layer light rugs.
Dust-free hay and bedding.
Follow veterinary instructions.
Do not pester. Only one person to attend to the horse if possible.

Q10.4

First signs of colic – vet.
Minor lameness – monitor.
Horse seen quidding – monitor.
Severe lameness – vet.
Minor graze to the quarters – basic treatment.
Suspected broken leg – vet.
Serious arterial bleed – vet.
Horse a little dull – monitor.

Q10.5

Medical history – veterinary treatments, ill-health and its duration.
Shoeing.
Vaccinations.
Dental treatment.
Chiropractic treatment.
Horse's normal TPR (temperature, pulse and respiration).
Horse's normal weight.

Q10.6

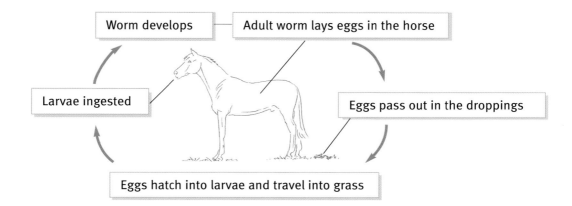

Worm develops — Adult worm lays eggs in the horse

Larvae ingested

Eggs pass out in the droppings

Eggs hatch into larvae and travel into grass

Q10.7

A horse-sick field,
Dull, lethargic.
Loss of appetite.
'Wormy' belly. The abdomen looks enlarged, whilst the rest of the horse looks poor.
Weight loss.
Cough.
Loss of performance.
Worms in droppings.
Yellow wax around anus.
Anaemia.
Dull coat.

Q10.8

(a) Conduct a worm count 3–4 times per year and worm accordingly. This is suggested by vets and wormer manufacturers, as worms are becoming resistant to wormers through simple routine worming.

(b) Routine worming around a yearly plan.
Base wormer, used as instructions year round except:
    February – small encysted.
    Summer – tapeworm.
    Dec/Jan– bots.
    December – small encysted.

Q10.9

Not accepting the contact when ridden; fussy in the mouth when ridden.
Quidding – the horse drops half-chewed food out of his mouth.
Takes a longer time to eat than normal.
Loss of appetite.
Bad smell from mouth.
Food stacked up in pouches in the sides of the mouth, like a hamster.

Q10.10

As the horse grinds his food, the outer edges of the upper jaw teeth and the inner edges of the lower jaw teeth are not worn down like the rest of the surface of the tooth because the upper jaw is slightly larger than the lower jaw. The dentist/vet rasps to make the teeth even.

11 FITTENING

Q11.1

(a) Check the horse's feet, and shoe accordingly.

Check vaccinations.

Check when the horse was last wormed.

Check dentist due date.

Check chiropractor due date.

Check that the tack fits and is good condition.

(b) Start to groom and trim up.

If the horse has been living out and is due to be stabled, start to bring the horse in from the field for short periods and give him a handful of hard feed, until spending the night in.

Q11.2

WEEK	1	2	3	4	5	6	7	8
WALK	15–30 min	30–60min	Yes	Yes	Yes	Yes	Yes	Yes
TROT	No	No	Short 5min trots	Build to more trot than walk	Yes	Yes	Yes	Yes
CANTER	No	No	No	No	Short 2–3min	Longer canters	Yes	Yes
LUNGEING	No	No	No	No	Yes	Yes	Yes	Yes
SCHOOLING	No	No	No	Light schooling	Yes	Yes possible dressage competition	Yes	
HACKING	Yes	Yes	Yes	Yes	Yes	Yes	Yes	Yes
HILLS	No	Yes	Yes (in walk)	Some trot hills	Walk and trot	Small hills in canter	Yes	Yes
JUMPING	No	No	No	No	Pole work	Start small jumps	Gridwork	Possible SJ comp.
FASTER CANTER WORK	No	No	No	No	No	Possibly	Pipe opener at end of week	Possibly

Q11.3

	MON	TUE	WED	THURS	FRI	SAT	SUN
WEEK 7	Lungeing ½ hr	Flatwork schooling ¾ hr Hack ½ hr	Hack 1hr WTC	Gridwork ¾ hr + hack ½ hr	Hack 1hr WTC	Pipe opener within hack	Day off
WEEK 8	Lungeing ½ hr	Flatwork schooling ¾ hr	Hack 1hr WTC + hills	Jump course ¾ hr	Hack 1hr WTC + hills	Dressage/ show jumping competition	Day off

Q11.4

	1	2	3	4	5	6	7	8
Fibre %	95	90	90	85	80	80	75	70–75
Concentrate %	5	10	10	15	20	20	25	30–25
Type of feed	Pony nuts	Pony nuts	Pony nuts	Pony nuts	If more energy required, oats, barley, alfalfa, sugar beet	Oats, alfalfa, barley, sugar beet	Oats, alfalfa, barley, sugar beet	Oats, alfalfa, barley, sugar beet

Q11.5

Change of environment from field to stable.

Change of feed – grass to hay.

Physical pressure – lungs working harder.

Stress.

Q11.6

Skin softened during time off.

Horse has changed shape and tack may not fit perfectly initially.

Tack may need attention if not well cared for.

Q11.7

RESPIRATORY	DIGESTIVE
• Cough • Nasal discharge	• Colic • Change in droppings, number and consistency • Scouring

Q11.8

WEEK	1	2	3
FEED	70% roughage, 30% concentrate, less heating concentrate	20% concentrate	10% concentrate down to none
WORK	Half workload	Light hacking, walk and trot	Light hacking walk
RUGS	1 rug less at night	2 rugs less at night	No rugs at night
SHOES			Shoes off end of week
TURNOUT	If not already, turn out am or pm	All day turnout	Out overnight, in short periods in the day

Q11.9

Sound. Warm enough. Eaten hay/feed. Signs of good health and maintaining weight.

Q11.10

CONCUSSION

Road work in trot.

Hard ground.

Jumping on a hard surface.

Fast work on hard ground.

STRAIN/SPRAIN

Deep going.

Jumping on a soft surface.

Fast work in deep going.

Trotting downhill.

Q11.11

Horse should be walked until breathing returns to normal – not blowing. This could be ridden, in hand or on a horse walker depending on circumstances and facilities.

Q11.12

Remove tack and clean later.

Wash as necessary – not cold water.

Return to stable, rug as necessary and leave to relax.

Groom and check the horse over. Look for areas of heat or swelling, particularly the legs.

12 GENERAL KNOWLEDGE

Q12.1

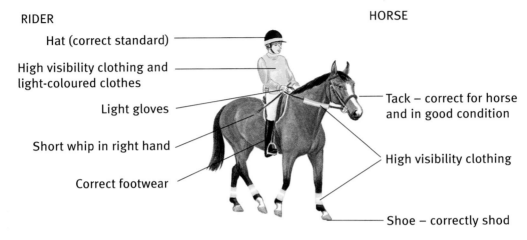

RIDER

- Hat (correct standard)
- High visibility clothing and light-coloured clothes
- Light gloves
- Short whip in right hand
- Correct footwear

HORSE

- Tack – correct for horse and in good condition
- High visibility clothing
- Shoe – correctly shod

Q12.2

Observe, signal, observe, manoeuvre.

Q12.3

1 – Observe, signal.

2 – Halt; look right, left, right; look behind right and if the situation has changed at all, signal again before manoeuvre.

3 – Walk a straight line to the other side of the road and turn right (looking both ways as you cross the road).

4 – Look in front and behind to assess new surroundings and hazards.

Q12.4

1 – Observe, signal.

2 – Observe behind right, lifesaver look left and turn left, if safe.

3 – Look down the new road for any hazards.

4 – Look in front and behind.

Q12.5

RIDER – Hat silk. Jacket. Tabard. Gloves.

HORSE – Bridle. Reins. Breastplate. Leg wraps. Exercise sheet. Saddle cover. Stirrup tabs.

Q12.6

Walk when passing other horses.

Leave gates as found.

Observe the Country Code.

Pass left to left in single file.

Q12.7

- Systematic training and education in all aspects of riding and horse management.
- Offer guidance and help throughout with breeding and training of horses and ponies.
- Actively encourage the protection of all horses and ponies.

Q12.8

Follow the APACH rules:

A = Assess

P = Prevent further accidents

A = Assess the casualty

CH = Call for Help

When assessing the situation, quickly observe all involved and how they have reacted to the accident. If other riders are nervous they may cause a further accident. To prevent this it may be necessary to clear the area or request help to hold horses and reassure riders. Ensure that the area is safe before approaching the casualty to assess their welfare. Ascertain if the casualty is conscious. If they are unconscious, send for an ambulance immediately, and if they are seriously bleeding, stem it with pressure. Carry out an ABC check – Airway, Breathing, Circulation. Ask the conscious casualty how they feel and where it hurts. Further help will come from the yard's qualified first-aiders.

13 GRASSLAND CARE

Q13.1

- Variety of quality grasses.
- Good grass density all over.
- All lawns, no roughs.
- No droppings – good picking up policy lessens the chance of worms and keeps grass sweet.
- Gently sloping land for free drainage.
- Good location for an automatic water trough.
- Well maintained post and rail fencing with electric fencing on the top to prevent chewing or leaning over the fence.
- Natural shelter from hedges on three sides, offering protection from sun, wind, rain.
- Free from poisonous plants.
- Gateway access 5-bar gate with padlocks both sides.

Q13.2

WINTER	SPRING	SUMMER	AUTUMN
Clear ditches	Harrow	Harrow	Harrow
Cut hedges	Seed	Fertilise (if necessary)	Roll if necessary and conditions allow
	Roll	Rotate stock or give some periods of rest	
	Fertilise. Check for poisonous plants	Keep checking for poisonous plants	

Q13.3

Cutting hedges encourages greater density. Hedges need to be kept back from fencing, especially electric, otherwise they can short the current, making it ineffective.

Q13.4

- Land, for poaching.
- Grass – length and density – is there enough to sustain the horses, or does the land need a rest?
- Water trough – functioning?
- Fencing – intact?
- Field shelter – safe?
- No poisonous plants.
- Horses – right number and all healthy.
- Droppings – daily or weekly removal.
- Litter and rubbish – daily removal.

Q13.5

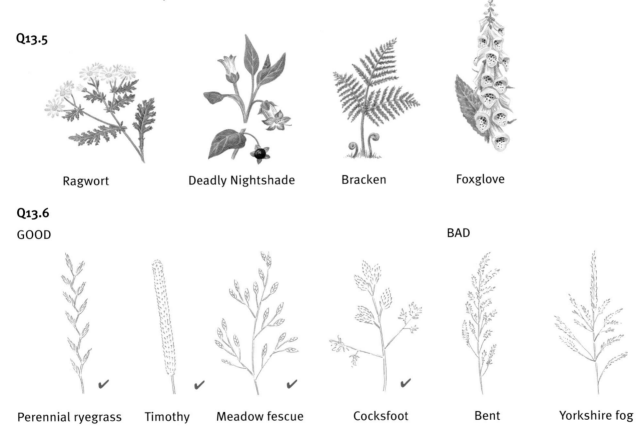

| Ragwort | Deadly Nightshade | Bracken | Foxglove |

Q13.6

GOOD   BAD

| Perennial ryegrass | Timothy | Meadow fescue | Cocksfoot | Bent | Yorkshire fog |

14 WATERING AND FEEDING

Q14.1

Continuous supply of fresh, clean water.

Water before feeding. Nowadays only a rule during cold weather when water may freeze. The rest of the time, generally horses have continuous access to water.

Keep water buckets clean and buckets deep enough for a good draught.

Never allow the horse to drink a large amount immediately after hard exercise when he is hot.

Do not work the horse straight after a large drink.

Q14.2

	ADVANTAGES	DISADVANTAGES
STREAM	Constant supply of water.	Can become poached and the horse stuck. A sand bed can give the horse sand colic. Toxic substances may accidentally contaminate the water.
BATH	Large, static vessel. Can see if the horses are drinking.	Remove taps and box in or the horse could be entangled or injured. Requires manual filling.
AUTOMATIC WATER TROUGH	Large vessel. Labour saving. Constant access.	Cannot tell if the horse is drinking.
BUCKETS	Can see if the horses are drinking.	Labour intensive. Small, therefore need many, and strong possibility of them getting knocked over.
WATER BUTTS	Can see if the horses are drinking.	Labour intensive. Preferable to buckets as they are larger and therefore less likely to get knocked over.

Q14.3

	RULE	REASON
AMOUNT	Feed the correct amount according to the horse's weight, work stabled/turnout, time of year, age, rider's ability, good doer/poor doer.	Maintain horse's weight, condition and ability to work.
QUALITY	Feed good quality fodder.	Poor quality feed is a false economy. The horse may need more or develop digestive or respiratory problems or may not eat.
HYGIENE	Use clean utensils.	Reduce the risk of infection and contamination and not put horse off food.
ROUTINE	Keep to the horse's feeding routine.	Changes in feeding routine can cause colic.
CHANGES	Make no sudden changes.	The gut needs to adapt slowly to changes in diet.
SUCCULENT	Variety and fresh vitamins.	Fresh source of vitamins and minerals and tempts fussy eaters.
FIBRE	Plenty of fibre.	Keeps the gut healthy and slows digestion so that all food is digested and absorbed efficiently and keeps digestive system functioning.
LITTLE AND OFTEN	Feed little and often. No more than 4–5lbs (1.8–2.2kg) of feed at any time.	The horse has a small stomach and naturally grazes 20 hrs per day. Feeding should mimic this.
EXERCISE	Allow 1 hour once the horse has eaten before exercising.	The blood supply will be in the gut, aiding digestion and cannot be used for muscular work. A full stomach can restrict lung capacity by pressing on the diaphragm.
WATER	Water before feeding.	Lots of water after feeding washes the feed through the system too quickly, possibly causing colic. (This rule is now rather outdated as most horses have free access to water.)

Q14.4

FEED	PROTEIN	CARBOHYDRATE	FATS/OIL	HEATING/FATTENING	FIBRE	FORM (e.g. flaked, bruised)	DESCRIPTION	MAXIMUM FEED	NOTES
OATS	AVERAGE	AVERAGE	AVERAGE	AVERAGE CAN BE HEATING	AVERAGE	ROLLED BRUISED NAKED CRUSHED CRIMPED	RICH YELLOW COLOUR SWEET SMELLING	N/A	LOW CALCIUM
BARLEY	AVERAGE	HIGH	AVERAGE	FATTENING CAN BE HEATING	LOW	FLAKED MICRONISED BOILED CRUSHED ROLLED	AS OATS WITH RIBS ALONG THE GRAIN	N/A	CAN HAVE ALLERGIC REACTION – BARLEY BUMPS
MAIZE	LOW	HIGH	LOW	HEATING FATTENING	LOW	FLAKED	LIKE CORN FLAKES	25% DAILY CONC	FOUND IN HEATING MIXES
BRAN	AVERAGE, BUT MOSTLY UNDIGESTIBLE	AVERAGE	LOW	NO	HIGH	FLAKES	FINE BROWN FLAKES SMELLS SOUR WHEN GONE OFF	N/A	WRONG CALCIUM: PHOSPHORUS RATIO
SUGAR BEET	LOW	HIGH	LOW	FATTENING	HIGH	PELLETS OR SHREDS	DARK PELLETS OR SHREDS	NOT TO FEED TO HORSES IN FAST WORK AS IT IS SLOW TO DIGEST	MUST BE SOAKED TO FEED
ALFALFA	HIGH + ESSENTIAL AMINO ACID LYSINE	AVERAGE	AVERAGE	AVERAGE	HIGH	SHORT CHOPPED FORAGE	GREEN CHOPPED PLANT, LUCERNE	N/A	
CHAFF	LOW	LOW	LOW	NO	HIGH	CHOPPED FORAGE	CHOPPED FORAGE	N/A	PREVENTS HORSE BOLTING FOOD
LINSEED	HIGH	HIGH	AVERAGE	FATTENING	LOW	SEEDS/OIL	SEEDS LOOK LIKE APPLE PIPS. OIL IS LIGHT YELLOW	3–4oz (85–113g) DRIED SEED 3 PER WEEK	SEEDS NEED COOKING OTHERWISE POISONOUS; LINSEED ADDS SHINE TO THE HORSE'S COAT
SOYA-BEAN MEAL	HIGH	HIGH	HIGH	FATTENING HEATING	LOW	MINCE	LIGHT COLOURED MINCE	LESS THAN 1lb (0.45kg) PER DAY	ONLY FOR HORSES IN HARD WORK
PEAS AND BEANS	HIGH	HIGH	LOW	HEATING FATTENING	LOW	CRUSHED OR SPLIT	GREEN BEANS	MAX 1lb (0.45kg) PER FEED	FOUND IN MIXES
MOLASSES	N/A	HIGH	N/A	FATTENING	N/A	DARK SYRUP	DARK SYRUP	SWEET TASTE TEMPTS FUSSY EATERS	CAN BE USED TO DISGUISE TASTE OF MEDICATION

Q14.5

Seed hay. Oats. Barley. Alfalfa. Maize. Peas and beans. Soya-bean meal.

Q14.6

	MAINTENANCE	LIGHT WORK
HAY/HAYLAGE	90%	80%
CONCENTRATE	10%	20%
TYPES OF CONCENTRATE	Pony nuts (compound) barley, alfalfa and sugar beet (straight)	Pony nuts (compound) barley, alfalfa and sugar beet (straight)

Q14.7

(a)

Digestive system less efficient.

Teeth may be less able to grind properly.

Generally lose weight as they get older.

Need higher level of protein.

(b)

Meadow hay – softer to chew.

Haylage – high nutritional value.

Wet feed – soaked barley rings, sugar beet.

Alfalfa – high in protein – older horses need higher levels of protein.

OAP mix – especially formulated for older horses.

Q14.8

Meadow hay. Sugar beet. Alfalfa. Convalescence mix. Succulents.

Q14.9

(a) PELLETS	(b) SHREDS	(c) SPEEDIBEET
Water 1:4	Water 1:2	Water 1:2
Soak 24 hours	Soak 12 hours	Soak 10 minutes

Q14.10

BRAN MASH

To 2lbs (0.9kg) bran, add enough boiling water to create a crumbly consistency. Add a little limestone flour to correct the calcium:phosphorus ratio. Cover and allow to cool until cold enough to eat. To check consistency, squeeze a handful of cooled bran into a ball. It should maintain the round form. Add sliced apples or carrots to make the feed more appealing.

LINSEED JELLY

Cover 2–3ozs (55–85g) of dried seed with water and soak overnight. Add 4 pints (1.8l) of water and simmer for 6 hours until all the seeds split. Allow to cool before feeding. To turn the jelly into a tea, add more water. Linseed is poisonous as dry seed and must be cooked before feeding.

Q14.11

A Fill and weigh the haynet.

B Put in bin and fill with water.

C Drain after 5–10 min, until no longer dripping.

D Tie haynet correctly in the stable.

Q14.12

Swells the hay seeds so that they are digested, not inhaled.

Washes spores away so that they are not inhaled.

Prevents horses coughing.

Q14.13

	HAY	HAYLAGE	SILAGE
SUITABLE FOR HORSES	Yes	Yes	No, but some feed if horses used to it; can be problematic
MOISTURE	15	60	80

Q14.14

	GALLEON am + lunch	GALLEON pm	LUCA am + lunch	LUCA pm	SIGI am + lunch	SIGI pm	DEXTER am + lunch	DEXTER pm	MELODY am	MELODY pm
HAYLAGE lbs (kg)	5 (2.2)	10 (4.5)	5 (2.2)	10 (4.5)	5 (2.2)	10 (4.5)	5 (2.2)	8 (3.6)	4 (1.8)	8 (3.6)
PONY NUTS lbs (kg)	3 (1.3)	4 (1.8)	3 (1.3)	4 (1.8)	3 (1.3)	4 (1.8)				
COMPETITION MIX – lbs (kg)							3 (1.3)	3 (1.3)		
ALFALFA lbs (kg)							1 (0.45)	1 (0.45)		
SUPPLEMENTS							10g electrolyte			
CARROTS									2	2

15 RIDING

Q15.1

Remove both feet from the stirrups. Pull the right buckle away from the stirrup bar. Cross the stirrup and leather over to lie neatly on the horse's left shoulder, in front of the saddle. Repeat with the left side.

Q15.2

Giving signals when riding on the road.
Riding and leading.

Q15.3

A whip is an object to inflict pain on the horse. FALSE
A whip is used only to make a horse go faster. FALSE
The whip should be used behind the leg to reinforce the leg aid. TRUE
The use of the whip should not interfere with the contact on the reins. TRUE
The whip should be changed over during the change of rein. FALSE

Q15.4

Consistency.
Repetition.
Positive reinforcement.
Patience.
Routine.
Honesty.

Q15.5

Pass left to left.
The outer track is for the faster paces.
Turn across the school for downward transitions.
The inner track is for the slower paces.
The outer track has right of way.
Lateral work has right of way.

Q15.6

HEAD – looking ahead
SHOULDERS – square and straight
BACK – straight
HIPS – out of saddle, soft joints
KNEES – lying on the saddle, soft joints
LOWER LEG – against the horse's side – security
ANKLES – soft joints
HEELS – down – security and balance

Q15.7

Use outer area of the field for the faster paces– that is, do not canter across the field.
To prevent accidents do not canter closely past another horse.
Use school figures to change the rein to maintain balance.

Q15.8

Downhill	Faster, less balanced.
Uphill	Slower, balanced.
Slippery, greasy	Unbalanced.
Heavy, deep	Slower, unbalanced.
Hard	Slower or faster, balanced.

Q15.9
To allow the rider to adopt and maintain a balanced jumping position.

Q15.10
The canter should be forward, balanced and rhythmical. During the turn on the approach to the fence, the quality of the canter is maintained by keeping the horse balanced between leg and hand. On the straight line approach, again the quality of the canter is maintained. The rider guides the horse to the centre of the fence and should look up and over the jump, focusing on the departure. As the horse jumps, the rider moves forward into jumping position, allowing the horse's head to move forward whilst maintaining a light contact at the end of the reins. The rider helps the horse to jump by remaining in balance with the horse. Maintaining a secure lower leg position allows the rider to land in balance and immediately rebalance the horse between leg and hand, recognise the canter lead and change it if necessary, using the next corner. The canter rhythm is re-established, if necessary, and a straight line ridden away from the fence, maintaining the quality of the canter through the departure turn.

Q15.11
Refusal.
Run out.
Poor jump.
Rider unbalanced over the fence.

Q15.12
RIDER:
Secure and balanced position.
Ability to use lower leg effectively.
Maintaining a consistent contact throughout.
Confidence.

HORSE:
Balanced.
Forward.
Rhythmical.
Good quality canter.
Desire to jump.

FURTHER READING

The following books and booklets can all be obtained from the BHS Bookshop (address overleaf).

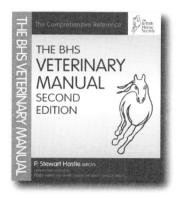

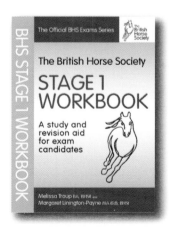

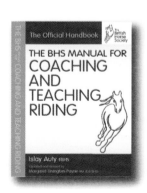
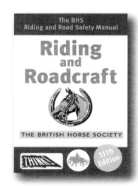

USEFUL ADDRESSES

The British Horse Society
Stoneleigh Deer Park
Kenilworth
Warwickshire
CV8 2XZ
tel: 08701 202244 or 01926 707700
fax: 01926 707800
website: www.bhs.org.uk
email: enquiry@bhs.org.uk

BHS Examinations Department
(address as above)
tel: 01926 707784
fax: 01926 707792
email:exams@bhs.org.uk

BHS Training Department
(address as above)
tel: 01926 707820
01926 707799
email: training@bhs.org.uk

BHS Riding Schools/Approvals Department
(address as above)
tel: 01926 707795
fax: 01926 707796
email: Riding.Schools@bhs.org.uk

BHS Bookshop
(address as above)
tel: 08701 201918
01926 707762
website: www.britishhorse.com